TELECOURSE GUIDE FOR

NUTRITION PATHWAYS CONCEPTS

NUTRITION: CONCEPTS AND CONTROVERSIES

A CONSUMER VIEW

Third Edition

Prepared by

Marie Yost Maness, Ph.D., Content Specialist,
Dallas County Community College District, Dallas, TX
Colleen Loveland, M.S., R.D., Telecourse Guide Developer

Produced by

Dallas TeleLearning
Dallas County Community College District

In association with
Wadsworth/Thomson Learning

© 2006 Dallas County Community College District.

All rights reserved. No part of this work may be reproduced, stored in a retrieval system, or transcribed, in any form or by any means — electronic, mechanical, photocopying, recording, or otherwise — without the prior written permission of the Dallas County Community College District.

Requests for permission to make copies of any part of the work should be mailed to:

Dallas TeleLearning
9596 Walnut Street
Dallas, Texas 75243

Printed in the United States of America
3 4 5 6 7 09 08 07

Printer: Thomson/West
ISBN-13: 978-0-495-01341-9
ISBN-10: 0-495-01341-2

For more information about our products, contact us at:
Thomson Learning Academic Resource Center
1-800-423-0563

For permission to use material from this text or product, submit a request online at
http://www.thomsonrights.com.
Any additional questions about permissions can be submitted by email to **thomsonrights@thomson.com.**

Thomson Higher Education
10 Davis Drive
Belmont, CA 94002-3098
USA

Contents

To the Student v
Telecourse Organization vii
Telecourse Guidelines ix

Lessons
1. Nutrition Basics 1
2. The Digestive System 13
3. Carbohydrates: Simple and Complex 23
4. Carbohydrates: Fiber 31
5. Fats: The Lipid Family 43
6. Fats: Health Effects 55
7. Protein: Form and Function 65
8. The Protein Continuum 79
9. Metabolism 89
10. Weight Control: Energy Regulation 97
11. Weight Control: Health Effects 107
12. Vitamins: Water Soluble 123
13. Vitamins: Fat Soluble 133
14. Major Minerals and Water 145
15. Trace Minerals 155
16. Physical Activity: Fitness Basics 165
17. Physical Activity: Beyond Fitness 175
18. Life Cycle: Pregnancy 183
19. Life Cycle: Lactation and Infancy 193
20. Life Cycle: Childhood and Adolescence 201
21. Life Cycle: Adulthood and Aging 211
22. Diet and Health: Cardiovascular Disease 219
23. Diet and Health: Cancer, Immunology, and AIDS 229
24. Diet and Health: Diabetes 239
25. Consumer Concerns and Food Safety 247
26. Applied Nutrition 261

Contributors 265

Personal Dietary Analysis Project 271

To the Student

To the Student

As a consumer, you may find that nutrition information can be confusing at times, even contradictory, leaving you with more questions than answers. Are there really "good" and "bad" foods? Is a low-carb, high-protein diet the best for long-term weight loss and health? If I eat fruit and starchy foods will I become fat? How much exercise do I need to lose weight and/or maintain weight loss? Should I take vitamin and mineral supplements as insurance? These and other questions will be answered based on nutrition science—and not on fads and fallacies.

As you explore *Nutrition Pathways—Concepts*, you will discover that the nutrition and lifestyle choices you make will either enhance or diminish your health and wellbeing. Embrace the learning experience and follow your personal Pathway to a healthier and more fulfilling destination.

—Marie Yost Maness, Ph.D.
Nutrition Content Specialist

Telecourse Organization

Nutrition Pathways-Concepts is designed as a comprehensive learning package in basic nutrition consisting of three elements: telecourse guide, textbook, and video programs. A computer-based activity component is available if your instructor chooses to offer it.

TELECOURSE GUIDE

The guide for this telecourse is:

Maness, Marie, and Colleen Loveland. *Telecourse Guide for Nutrition Pathways-Concepts.* 3rd ed. Belmont, CA: Wadsworth/Thomson Learning, 2006. ISBN: 0-495-01341-2

The telecourse guide acts as your daily instructor. If you follow the Telecourse Guidelines carefully, you should successfully accomplish all of the requirements for this course. (See the section entitled Telecourse Guidelines.)

TEXTBOOK

In addition to the telecourse guide, the textbook required for this course is:

Sizer, Frances, and Ellie Whitney. *Nutrition: Concepts and Controversies.* 10th ed. Belmont, CA: Wadsworth/Thomson Learning, 2006. ISBN: 0-534-64506-2

VIDEO PROGRAMS

The video program series for this telecourse is:

Nutrition Pathways

Each video program is correlated with the telecourse guide and the reading assignment for that lesson. Be sure to read the Lesson Focus Points in the telecourse guide before you watch the program. The video programs are presented in a documentary format and are designed to bring analysis and perspective to the issues being discussed. Watch them closely.

If the programs are broadcast more than once in your area, or if video or audio tapes are available at your college, you might find it helpful to watch the video programs more than once or to listen to an audio tape for review. Since examination questions will be taken from the video programs as well as from the reading, careful attention to both is vital to your success.

INTERACTIVE COURSE

Self-graded interactive exercises, pre- and post-self-assessments, and case-based, problem-solving scenarios are available to students whose institutions have opted to offer these. These activities are useful for reinforcement and review of lesson content and learning objectives. The interactive activities are offered in two formats: CD-ROM and Internet. Ask your instructor how to access these activities if they are listed in your syllabus as a course requirement.

Telecourse Guidelines

Follow these guidelines as you study the material presented in each lesson:

1. OVERVIEW—
   Read the Overview for an introduction to the lesson material.

2. LEARNING OBJECTIVES—
   Review the Learning Objectives and pay particular attention to the lesson material that relates to them.

3. LESSON ASSIGNMENT—
   Review the lesson Assignment in order to schedule your time appropriately. Pay careful attention; the titles and numbers of the textbook chapter, the telecourse guide lesson, and the video program may be different.

4. LESSON FOCUS POINTS—
   Get the most from your reading your textbook and watching the video by reviewing the Lesson Focus Points in each lesson before you read your textbook assignment and watch the lesson video.

5. KEY CONCEPTS—
   The purpose of this part of your telecourse guide is to point out and clarify essential concepts and principles contained in this lesson. By reviewing this material and relating it to that in the textbook, you should be able to increase your understanding and ability to successfully complete homework assignments. (All lessons, with the exception of Lessons 14, 15, 19, 21, and 26 include Key Concepts.)

6. RELATED ACTIVITIES—
   The Related Activities are not required unless your instructor assigns them. They are offered as suggestions to help you learn more about the material presented in the lesson.

7. PRACTICE TEST—
   Complete the Practice Test to help you evaluate your understanding of the lesson.

8. ANSWER KEY—
Use the Answer Key at the end of the lesson to check your answers or to locate material related to each question of the Practice Test.

9. PROJECT—
The Personal Dietary Analysis Project is designed to provide you with specific information about nutrient adequacy and balance, number of kcalories consumed, nutrient density, and variety of the foods you typically eat. The project forms are located at the back of the telecourse guide, starting on page 271.

Diet Analysis Plus software is available from the textbook publisher. Input your food records using the software to generate the required printouts. The users' guide for the computerized dietary analysis program has samples of what you will need to print.

Lesson 1

Nutrition Basics

OVERVIEW

Whether we realize it or not, nutrition plays a major role in our lives—from before we are born and into our golden years. This lesson will demonstrate how the daily food choices people make throughout their lives can influence how well they live and possibly how long they will be around. Choosing more nutrient-dense foods or less nutrient-dense, higher calorie foods will have an effect on health—either positive or negative. You will learn how the development of nutrition as a science has affected what we eat and how food choices reflect your family traditions, religious beliefs, emotions, financial status, socialization, food preferences, and, of course, nourishment. How do you go about making reasonable choices from the many foods available to you? Do you select foods based solely on nutritional value and disregard the other reasons people choose foods? These questions and others will be answered in Lesson 1.

In this lesson you will also be introduced to diet planning principles and guidelines. These guidelines will help you select foods that provide energy without excess calories and foods that provide variety, balance, and nutrient adequacy.

LEARNING OBJECTIVES

Upon completing this lesson, you should be able to:

1. Describe nutritional genomics.
2. Describe factors which affect food choices.
3. Describe the six major classes of nutrients.
4. Describe the process by which scientists uncover nutrition facts.
5. Distinguish between valid sources of nutrition information and nutrition misinformation.

6. Describe the roles of Dietary Reference Intakes (DRI).
7. Cite the Dietary Guidelines for Americans.
8. Explain how foods are grouped in the Food Guide Pyramid.
9. Identify the advantages and disadvantages of the Food Guide Pyramid.
10. Explain the function of the Exchange Lists.
11. Describe the components of a food label.
12. Describe phytochemical and possible health benefits.
13. Identify the health concern for each person in the *Nutrition Basics* video.

LESSON ASSIGNMENT

Review the following assignment in order to schedule your time appropriately. Pay careful attention; the titles and numbers of the textbook chapter, the telecourse guide lesson, and the video program may be different from one another.

Text: Sizer and Whitney, *Nutrition: Concepts and Controversies*, Chapter 1, "Food Choices and Human Health," pp. 1–26; Chapter 2, "Nutrition Tools—Standards and Guidelines," pp. 27–64

Video: "Nutrition Basics" from the series *Nutrition Pathways*

LESSON FOCUS POINTS

The following questions are designed to help you get the most out of the lesson. After you read the textbook assignments and view the video, you should be able to answer these questions.

1. What is nutritional genomics? How will nutritional genomics impact food choices? (text, p. 4; video)

2. What are the six classes of nutrients? Define the terms *inorganic* and *organic*. What makes a nutrient "essential"? What are the energy-yielding nutrients? Which nutrients do not yield energy when broken down? How many calories are in each energy-yielding nutrient? Define *kcalorie* and *nutrient density*. What are elemental diets? What are phytochemicals? What are the five characteristics of a nutritious diet? (text, pp. 4–8; video)

3. What are the various factors that affect our food choices? (text, p. 10; video)

4. Why is nutrition a science? Name four types of research studies and the purpose of each type of study. What is the "placebo effect"? (text, pp. 11–13)

5. What is the main goal of the Dietary Guidelines for Americans? (text, p. 14)

6. Where can you find reliable nutrition information? (text, pp. 24–26)

7. What is the DRI? What do the DRI values include? Define Estimated Average Requirement, Recommended Dietary Allowances, Adequate Intakes, and Tolerable Upper Intake Level. What are the four goals of the DRI? (text, pp. 28–30)

8. What are the differences between the following food group plans: the U.S. Department of Agriculture (USDA) Food Guide 2005 and ethnic food guides. (text, pp. 35–36; 42–43)

9. What foods are represented on each level of the pyramid? What should be emphasized when selecting foods from the pyramid? How much is in a serving of the following foods: pasta, raw vegetables, fruit, milk, and meat? (text, pp. 41, 44; video)

10. What are the advantages and disadvantages in using the Food Guide Pyramid? (text, p. 41)

11. How is food sorted in exchange lists? Why were exchange lists originally developed? (text, p. 45; video)

12. What information should every food label display? In what order are ingredients given on a food label? What are the Daily Values? What relationships for health claims have been approved by the Food and Drug Administration (FDA) for use on food labels? How can a food be disqualified from making health claims on labels? (text, pp. 47–50)

13. Define phytochemicals and functional foods. What are the recommendations for consuming phytochemicals? (text, pp. 57, 64)

14. What was developed by nutrition scientists during the 1940s to help the nation improve nutrient status and prevent deficiencies? (video)

15. What nutrient provides us with quick energy and is the main fuel source for the brain? What nutrient is considered the "building block" of life? What nutrient provides us with concentrated energy and is necessary for many cell functions? What nutrient helps to activate chemical reactions in the body? What class of nutrients helps to build strong bones and teeth? What nutrient is necessary to support every chemical reaction in the body? (video)

16. Describe briefly the health concerns of each Pathway subject who will be followed for one year. (video)

KEY CONCEPTS

Some Reasons for Food Choices

- Personal preference
- Habit
- Ethnic heritage
- Social interactions
- Availability, convenience, economics
- Positive or negative association
- Emotional comfort
- Values (religious, environmental, political)
- Body image
- Nutrition

Exchange Lists

- Originally developed for use with diabetics
- Foods grouped according to protein, carbohydrate, and fat content
- Can be used for weight loss

The USDA Food Guide Pyramid

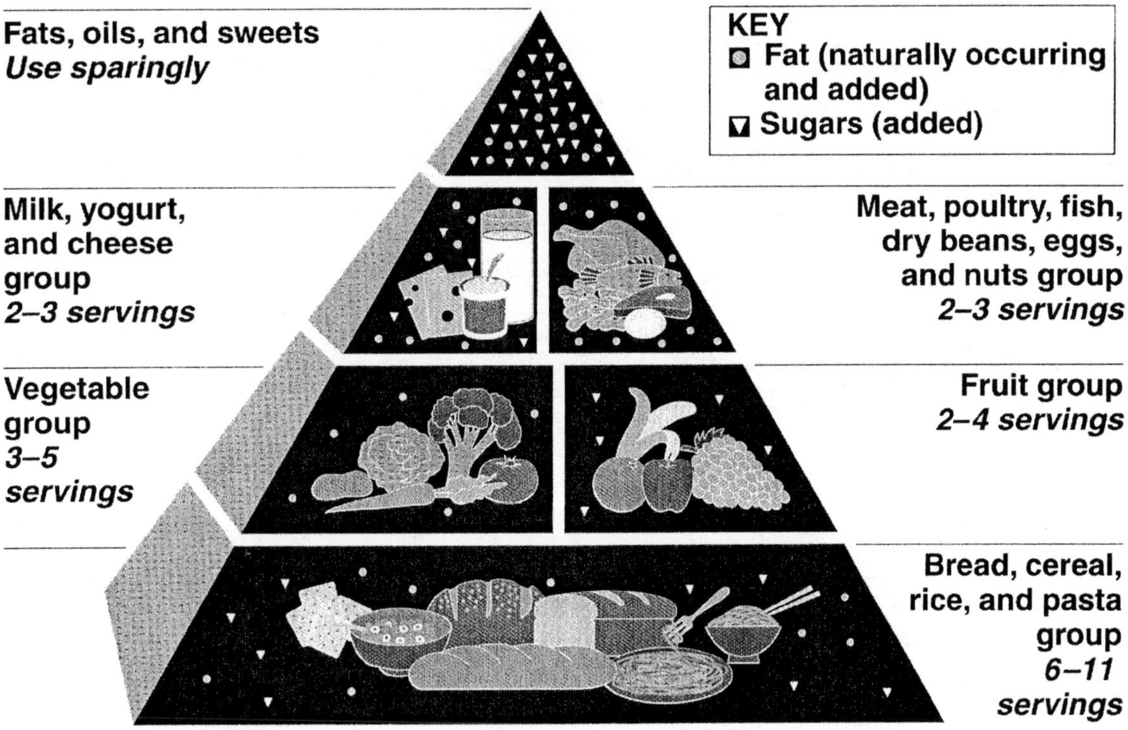

© 2000 Wadsworth/Thomson Learning

RELATED ACTIVITIES

These activities are not required unless your instructor assigns them. They are offered as suggestions to help you learn more about the material presented in this lesson.

- ❑ Compare Your Food Intake to the Daily Food Guide Pyramid—Complete the form on the following page and return it to your instructor according to the established deadline.
- ❑ Tufts University Nutrition Navigator

Compare Your Food Intake to the Daily Food Guide Pyramid

Record what you ate yesterday. Indicate the number of servings you consumed from each food group and tally up the totals. Look at the previous page to see how your food intake compares to the suggested daily food intake on the Food Guide Pyramid. See pages 38–40 of your textbook for serving sizes in each food group.

List food and amount	Bread, cereal, rice, and pasta	Fruit	Vegetable	Meat, poultry, fish, dry beans, eggs, and nuts	Milk, yogurt, and cheese	Misc.
Breakfast: __________ __________ __________ __________						
Snack: __________ __________ __________ __________						
Lunch: __________ __________ __________ __________						
Dinner: __________ __________ __________ __________						
Snack: __________ __________ __________ __________						
Totals						

Source: Teaching Concept by Simin Bolourchi Vaghefi, Ph.D., College of Health, University of North Florida

Tufts University Nutrition Navigator

The purpose of this assignment is to help familiarize you with the *Tufts University Nutrition Navigator*, an online rating and review guide. The *Tufts University Nutrition Navigator* is designed to help with identifying reliable and accurate sources of nutrition information. Websites are reviewed by Tufts nutritionists, who apply rating and evaluation criteria developed by the Tufts University Navigation Advisory Board, a prestigious panel of leading U.S. and Canadian nutrition experts. Site reviews are updated quarterly to ensure that ratings take into account the ever changing Internet and nutrition environments.

1. Go to any library or use your home computer and get online.
2. Enter this location address: http: //navigator.tufts.edu/
   Be sure to carefully enter the address; pay close attention to spelling and punctuation, and do not capitalize any of the letters.
3. If you have correctly entered the above address and you go online, then you will find the *Tufts University Nutrition Navigator* home page displayed on your monitor.
4. Select one of the generalized topic areas displayed on your screen: Educators, General Nutrition, Journalists, Kids, Health Professionals, Special Dietary Needs, Parents, Women. Begin your search with one of these selections. Alternatively, you may search for information on another specific topic that interests you such as: antioxidants, vitamin B_{12}, osteoporosis, HIV, cancer, etc.
5. Describe your findings on the topic you researched in a typed, or neatly handwritten, one- to two-page summary.
6. Print a review of a website or a home page from a website from one of the above topic areas listed in #4.
7. Read the review or home page.
8. Turn in both the website review or home page stapled to the summary.

Source: Marvin Parent, Oakland Community College

PRACTICE TEST

The following items will help you evaluate your understanding of this lesson. Use the answer key at the end of the lesson to check your answers or to locate material related to each question.

Multiple Choice

Select the one choice that best answers the question.

1. Which of the following represents a food choice based on positive association?
   A. A tourist from China who rejects a hamburger due to unfamiliarity
   B. A child who spits out his mashed potatoes because they taste too salty
   C. A teenager who grudgingly accepts an offer for an ice cream cone to avoid offending a close friend
   D. An elderly gentleman who eats a peanut butter and jelly sandwich because it reminds him of his happy childhood

2. Which of the following nutrients does NOT yield energy during its metabolism?
   A. Fat
   B. Protein
   C. Vitamins
   D. Carbohydrates

3. Which of the following provides the most energy?
   A. Fats
   B. Alcohol
   C. Proteins
   D. Carbohydrates

4. What is the calorie value of a meal supplying 110 grams of carbohydrates, 25 grams of protein, 20 grams of fat, and 5 grams of alcohol?
   A. 160
   B. 345
   C. 580
   D. 755

5. In nutrition, the word *essential* means __________.
   A. necessary for good health and proper functioning of the body
   B. a necessary nutrient that can be obtained only from the diet
   C. that the body can manufacture the nutrient from raw materials
   D. compounds the body can make for itself

6. Which of the following is NOT a set of values within the Dietary Reference Intakes?
   A. Adequate Intake
   B. Estimated Average Intake
   C. Tolerable Upper Intake Level
   D. Recommended Dietary Allowance

7. Which of the following statements is NOT true about the Dietary Reference Intakes (DRI)?
   A. The committee that publishes them is appointed by the government.
   B. They are requirements, not recommendations.
   C. They are based on review of available scientific research.
   D. They are for individuals who are healthy.

8. Of the following, the MOST valid source of nutrition information are ________.
   A. newspaper articles
   B. TV programs
   C. scientific journals
   D. health magazines

9. Which of the following is NOT one of the Dietary Guidelines for Americans?
   A. Choose a diet with plenty of grain products.
   B. Choose a diet high in sugar.
   C. Choose a diet moderate in salt.
   D. Eat a variety of foods.

10. The major drawback of the Food Guide Pyramid is that __________.
   A. it is more expensive to follow than other plans
   B. it does not limit food choices to foods low in calories
   C. it is not simple to use
   D. it makes shopping more difficult and time consuming

11. In the Food Guide Pyramid, the small dots and triangles scattered in the background represent __________.
A. fats and sugars
B. salt and pepper
C. calcium and iron
D. vitamins and minerals

12. Exchange lists help primarily with __________.
A. balance
B. calorie control
C. adequacy
D. moderation

13. According to nutrition labeling laws, the amounts of what two vitamins must be listed on the package label?
A. Vitamins D and E
B. Vitamins A and C
C. Thiamin and riboflavin
D. Vitamin B_6 and niacin

14. One of the major disadvantages of the Food Guide Pyramid is that it cannot be adapted to other national and cultural cuisine.
A. True
B. False

15. Scientists' responsibilities include all of the following EXCEPT __________.
A. obtaining facts by systematically asking questions
B. conducting experiments to test for possible answers
C. submitting their findings to the news media
D. publishing the results of their work in scientific journals

16. Proteins are considered the __________.
A. main source of quick energy for the body
B. building blocks of life
C. activator of chemical reactions in the body
D. best source of concentrated energy

17. Nearly every chemical reaction in the body is activated by __________.
    A. fat
    B. carbohydrates
    C. vitamins
    D. water

18. What foods are found on the base of the Food Guide Pyramid?
    A. Fruits and vegetables
    B. Grains and cereals
    C. Dairy products
    D. Meat products

Fill in the Blank

Insert the correct word or words in the blank for each item.

19. According to the Food Guide Pyramid, ____________ constitutes a serving of milk.

20. The three health concerns of the Pathway subjects being profiled for one year are ____________, ____________, and ____________.

Essay Questions

Reflect on the items listed below; one or more may be included in the examination.

21. Explain how food choices are influenced by habits, emotions, physical appearance, and ethnic background.

22. What are the major advantages and drawbacks of the Food Guide Pyramid?

ANSWER KEY

The following provides the answers and references for the practice test questions.

	Answer	Learning Objectives	References
1.	D	LO 2	Sizer/Whitney, p. 10
2.	C	LO 3	Sizer/Whitney, p. 5
3.	A	LO 3	Sizer/Whitney, p. 5
4.	D	LO 3	Sizer/Whitney, p. 6
5.	B	LO 3	Sizer/Whitney, p. 5
6.	B	LO 6	Sizer/Whitney, p. 29
7.	B	LO 6	Sizer/Whitney, p. 31
8.	C	LO 4	Sizer/Whitney, p. 14
9.	B	LO 7	Sizer/Whitney, p. 16
10.	B	LO 8	Sizer/Whitney, p. 41
11.	A	LO 8	Sizer/Whitney, p. 41
12.	B	LO 10	Sizer/Whitney, p. 45
13.	B	LO 11	Sizer/Whitney, p. 47
14.	B	LO 9	Sizer/Whitney, pp. 41–42
15.	C	LO 5	Sizer/Whitney, p. 11; Video
16.	B	LO 3	Video
17.	C	LO 3	Video
18.	B	LO 8	Video
19.	one cup of milk	LO 8	Video
20.	Type 2 diabetes (diabetes), high blood cholesterol (CVD, heart disease), and weight loss	LO 13	Video
21.		LO 2	Sizer/Whitney, p. 10
22.		LO 9	Sizer/Whitney, pp. 41–42

Lesson 2

The Digestive System

OVERVIEW

This introduction to digestion will examine the common pathways that all foods take—from the time you put food in your mouth, to the transformation of foods into nutrients, to the transport of absorbed nutrients to the cells of the body, to the final elimination of the unused nutrients and/or waste products. During the journey, you will observe many problems encountered during digestion and watch how the body miraculously solves those problems.

LEARNING OBJECTIVES

Upon completing this lesson, you should be able to:

1. Describe the major functions of the cardiovascular, hormonal, nervous, digestive, and excretory systems.
2. Identify the basic route followed by food through the gastrointestinal (GI) tract and the main function of each organ.
3. Describe the mechanical processes of digestion in order of their occurrence in the body.
4. Describe the chemical processes of digestion in order of their occurrence in the body.
5. Identify the unique structure of the small intestine which enhances absorption.
6. Name the two transportation routes that absorbed nutrients take in order to be delivered and used by the body.
7. Identify common digestive problems, explain how the GI tract is involved, and list common solutions to the problems.
8. Describe alcohol's effect on nutrition.

LESSON ASSIGNMENT

Review the following assignment in order to schedule your time appropriately. Pay careful attention; the titles and numbers of the textbook chapter, the telecourse guide lesson, and the video program may be different from one another.

Text: Sizer and Whitney, *Nutrition: Concepts and Controversies*, Chapter 3, "The Remarkable Body," pp. 65–98

Video: "The Digestive System" from the series *Nutrition Pathways*

LESSON FOCUS POINTS

The following questions are designed to help you get the most out of the lesson. After you read the textbook assignments and view the video, you should be able to answer these questions.

1. Which nutrients are transported directly through the vascular (bloodstream) system? Which nutrients are transported through the lymphatic system and how do they enter the blood circulatory system? (text, p. 68)

2. What are the four main body systems that interact with the digestive system? What are the major functions of each of these systems? (text, pp. 68–74)

3. Where does the process of digestion begin? What are the functions of the mouth, the esophagus, the stomach, the small intestine, the large intestine, the rectum, and the anus? (text, pp. 76–77; video)

4. What is peristalsis and where does it begin? What is the function of a sphincter muscle? What is chyme? (text, pp. 77–78; video)

5. What nutrient is partially digested by enzymes contained in saliva? Why is the presence of hydrochloric acid necessary in the stomach? What is gastric juice? What is the function of mucus? What nutrient is partially broken down in the stomach? What organ produces bile? Where is bile stored? What is the function of bile? What is pancreatic juice? Why is the presence of sodium bicarbonate necessary in the small intestine? What is the function of fiber in digestion? (text, pp. 79, 81)

6. Where does the primary absorption of nutrients take place in the body? What are villi? What are microvilli? What is the main function of villi and microvilli? (text, p. 83)

7. What is heartburn/acid indigestion? What could help alleviate the symptoms of heartburn/acid indigestion? What danger is associated with vomiting? (text, pp. 85–86; video)

8. What is the function of belching and what foods may cause an increase in belching? What causes flatus and what distinguishes it from bloating? (text, pp. 85–86; video)

9. What is constipation? How can constipation be prevented or treated? (text, p. 86; video)

10. What is the cause of peptic ulcers? (text, p. 86; video)

11. What are the signs of choking? What technique could you apply to assist a choking person? (text, pp. 86–87; video)

12. What impact does alcohol have on nutrition? What impact does alcohol have on body fat? Is there any health benefit to alcohol consumption and heart disease? (text, pp. 96–98)

The Gastrointestinal Tract

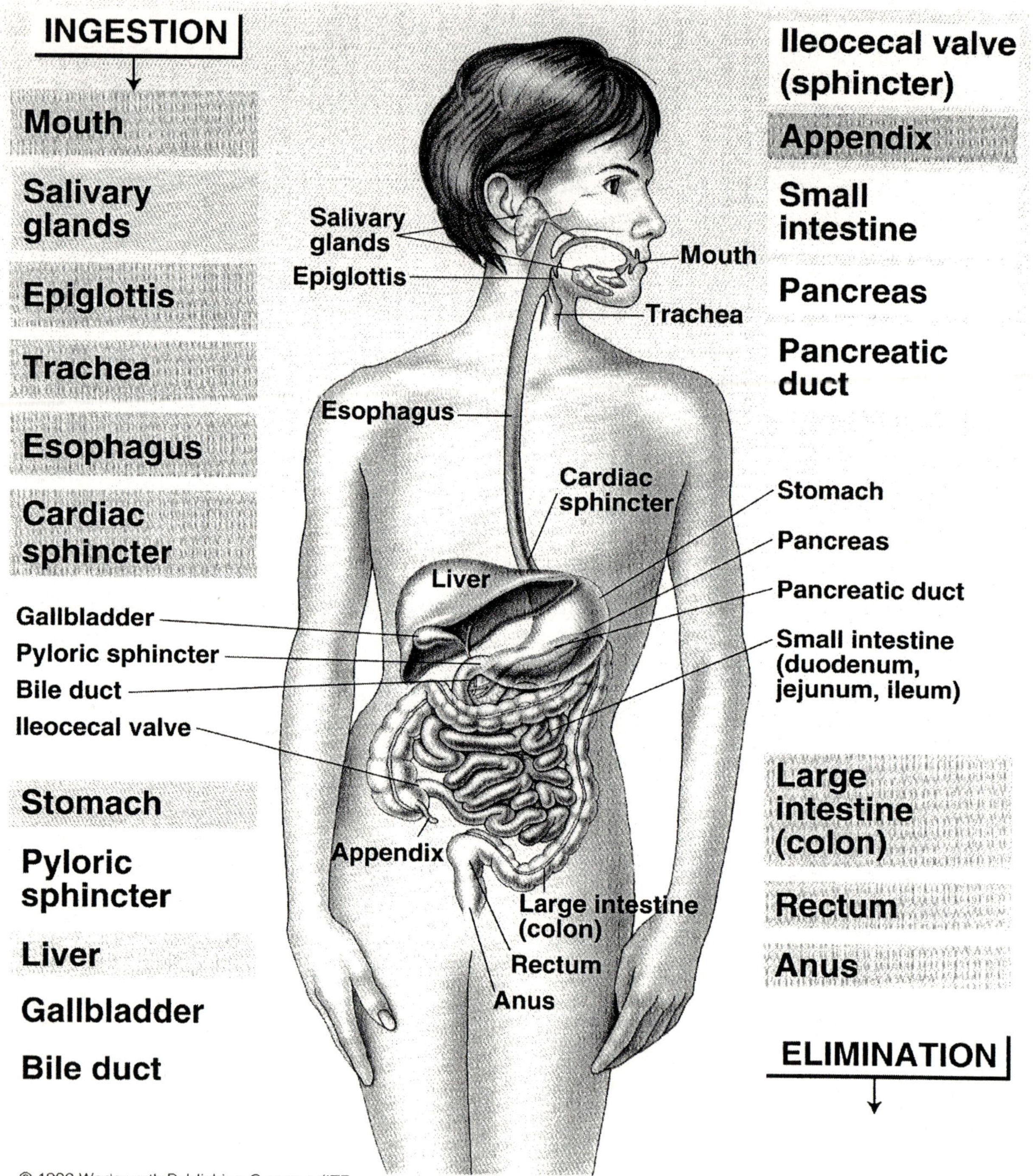

© 1999 Wadsworth Publishing Company/ITP

Summary of Digestion

Food	Agents (site)	Products
I. Carbohydrate: starch and sugar	Carbohydr*ase* → (mouth, small intestine)	glucose
II. Protein: animal and vegetable protein	Hydrochloric acid (HCl), Prote*ase* → (stomach, small intestine)	amino acids
III. Fat (lipid): animal fat and vegetable oil	Bile, Lip*ase* → (small intestine)	fatty acids + glycerol

Digestive Juices and Their Contents

Site of Secretion	Ph	Contents
Salivary glands in mouth: Saliva	Basic	Carbohydrase
Stomach: Gastric juice	Acidic	Hydrochloric acid (HCl) Protease
Liver and gall bladder: Bile	Basic	Bile
Pancreas: Pancreatic juice	Basic	Carbohydrase Protease Lipase Sodium bicarbonate
Small intestine: Intestinal enzymes	Basic	Protease Sucrase Maltase Lactase

PRACTICE TEST

The following items will help you evaluate your understanding of this lesson. Use the answer key at the end of the lesson to check your answers or to locate material related in each question.

Multiple Choice

Select the one choice that best answers the question.

1. The primary organ of digestion and absorption is the __________.
   A. mouth
   B. stomach
   C. small intestine
   D. large intestine

2. Which of the following is a function of sphincter muscles?
   A. Control peristalsis
   B. Grind large food particles
   C. Secrete digestive juices into the GI tract
   D. Control the passage of food through the GI tract

3. What is a function of gastric juice in the stomach?
   A. Absorbs water
   B. Stops peristalsis
   C. Neutralizes the food mass
   D. Makes it an acidic environment

4. What is the name of the projections on the inner surface of the small intestine?
   A. Villi
   B. Cilia
   C. Blood vessels
   D. Capillaries

5. Without sufficient __________, the digestive tract muscles get too little exercise and become weak.
   A. energy
   B. fiber
   C. nutrients
   D. water

6. You have just consumed a meal very high in fat. As a result, hormonal messages will tell an organ to send __________ in amounts matched to the amount of fat present.
   A. bile
   B. bicarbonate
   C. hydrochloric acid
   D. mucus

7. Which of the following characterizes gas and bloating?
   A. Gas is a substance.
   B. Gas can be measured.
   C. Bloating is a sensation.
   D. All of the above.

8. The use of an antacid is indicated PRIMARILY for which condition?
   A. Excessive gas
   B. Acid indigestion
   C. Excessive belching
   D. Bloating

9. Therapy for constipation would include ALL of the following EXCEPT _____.
   A. increasing water intake
   B. decreasing fiber intake
   C. increasing physical activity
   D. responding promptly to the defecation signal

Fill in the Blank

Insert the correct word or words in the blank for each item.

10. The most obvious sign of choking is that the person cannot ______________.

11. The function of belching is to rid the small intestine or stomach of ______________.

12. The greatest danger from vomiting is ________________.

13. The two transportation routes that absorbed nutrients take in order to be delivered and used by the body are ___________ and ___________.

Matching

Select the letter next to the word or phrase that best matches the numbered statements. Use answers only once.

Match the digestive organs, listed on the right, with their appropriate functions, listed on the left.

_____	14. Manufactures bile to help digest fats	A. Mouth
_____	15. Conducts bile into the small intestine	B. Esophagus
_____	16. Opens to allow elimination	C. Stomach
_____	17. Churns, mixes, and grinds food to a liquid mass	D. Small intestine
		E. Liver
_____	18. Reabsorbs water and minerals	F. Gallbladder
_____	19. Stores bile until needed	G. Pancreas
_____	20. Passes food to stomach	H. Large intestine
_____	21. Manufactures enzymes to digest the three energy-yielding nutrients; releases sodium bicarbonate	I. Rectum
		J. Anus
_____	22. Chews and mixes foods with saliva	K. Bile duct
_____	23. Stores waste prior to elimination	L. Pancreatic duct
_____	24. Secretes enzymes that digest carbohydrate, fat, protein; absorbs nutrients	

Essay Questions

Reflect on the items listed below; one or more may be included on the exam.

25. Describe how hormones affect nutrition.

26. Describe four common digestive problems and their recommended treatments or therapies.

ANSWER KEY

The following provides the answers and references for the practice test questions.

	Answer	Learning Objectives	References
1.	C	LO 2, 5	Sizer/Whitney, p. 78
2.	D	LO 3	Sizer/Whitney, p. 77
3.	D	LO 4	Sizer/Whitney, p. 79
4.	A	LO 5	Sizer/Whitney, p. 83
5.	B	LO 3	Sizer/Whitney, p. 80
6.	A	LO 1, 4	Sizer/Whitney, p. 79
7.	D	LO 8	Video
8.	B	LO 8	Video
9.	B	LO 8	Video
10.	speak (make a sound, cough)	LO 8	Video
11.	carbon dioxide (gas)	LO 8	Video
12.	dehydration	LO 8	Video
13.	blood, lymph	LO 6	Sizer/Whitney, p. 83
14.	E	LO 8	Sizer/Whitney, p. 76
15.	K	LO 4	Sizer/Whitney, p. 76
16.	J	LO 2	Sizer/Whitney, p. 76
17.	C	LO 2, 3	Sizer/Whitney, p. 76
18.	H	LO 2	Sizer/Whitney, p. 76
19.	F	LO 4	Sizer/Whitney, p. 76
20.	B	LO 2	Sizer/Whitney, p. 76
21.	G	LO 4	Sizer/Whitney, p. 76
22.	A	LO 3, 4	Sizer/Whitney, p. 76
23.	I	LO 2	Sizer/Whitney, p. 76
24.	D	LO 2, 4	Sizer/Whitney, p. 76
25.		LO 1	Sizer/Whitney, pp. 71–72
26.		LO 8	Sizer/Whitney, pp. 85–86

Lesson 3

Carbohydrates: Simple and Complex

OVERVIEW

In this lesson, the carbohydrate family from sugar to starch will be examined. The complex carbohydrate, fiber, will be the topic of another lesson. You will see how dietary carbohydrates are digested and absorbed, converted to glucose for immediate energy, and converted to glycogen for energy storage. You will discover how blood glucose is regulated within the body and how insulin and glucagon work to maintain a normal blood glucose level. The health effects of carbohydrates will also be examined as well as the health implications of lactose intolerance and refined sugar intake.

LEARNING OBJECTIVES

Upon completing this lesson, you should be able to:

1. Identify the primary fuel source for the brain and nervous system.
2. Define *simple carbohydrates* and identify foods which contain simple carbohydrates.
3. Define *complex carbohydrates* and identify foods which contain complex carbohydrates.
4. Cite the recommended intake of carbohydrates.
5. Describe the basic steps involved in the digestion and absorption of starch.
6. Define *lactose intolerance* and its health effects.
7. Describe the body's use of glucose to provide energy or to make glycogen.
8. Identify the hormones involved in blood glucose regulation.
9. Distinguish between fasting and postprandial hypoglycemia.
10. Describe the health effects of sugar and cite the recommended intake of sugar.

LESSON ASSIGNMENT

Review the following assignment in order to schedule your time appropriately. Pay careful attention; the titles and numbers of the textbook chapter, the telecourse guide lesson, and the video program may be different from one another.

Text: Sizer and Whitney, *Nutrition: Concepts and Controversies*, Chapter 4, "The Carbohydrates: Sugar, Starch, Glycogen, and Fiber," pp. 99–138

Video: "Carbohydrates: Simple and Complex" from the series *Nutrition Pathways*

LESSON FOCUS POINTS

The following questions are designed to help you get the most out of the lesson. After you read the textbook assignments and view the video, you should be able to answer these questions.

1. What are simple carbohydrates? What foods provide simple carbohydrates? What are the three monosaccharides? What monosaccharide is the sweetest? What foods contain fructose? What monosaccharide is found in milk? What foods contain galactose? What does the term *disaccharide* mean? Name the disaccharides. What is the common component of all disaccharides? What disaccharide is also called table sugar? What is the disaccharide found in milk sugar? Which disaccharide is formed mainly during the breakdown of starch during digestion? (text, pp. 100–101; video)

2. What simple carbohydrate is the primary fuel source for the brain? (text, p. 101; video)

3. What does the term *polysaccharide* mean? What foods provide starch carbohydrates? What two dietary complex carbohydrates are important in nutrition? What are the richest food sources of starch? (Reminder: Fiber will be the focus of Lesson 4.) (text, p. 102)

4. Describe the health effects of complex carbohydrates in the diet. What is the recommended intake of carbohydrate as a percentage of total energy? (text, p. 105)

5. What is the ultimate goal of carbohydrate digestion and absorption? Where does starch digestion begin? Where does most carbohydrate digestion occur? Name the enzymes that dismantle the following disaccharides: maltose, sucrose, and lactose. Where does absorption of monosaccharides occur? (text, pp. 113–114)

6. What monosaccharide is the primary fuel for the brain? (text, p. 114)

7. What enzyme is absent or deficient in people suffering from lactose intolerance? What are the symptoms and causes of lactose intolerance? What populations have the lowest and highest prevalence of lactose intolerance? (text, p. 116; video)

8. What is the primary role of carbohydrates in human nutrition? What is the function of the liver in carbohydrate metabolism? What is the storage form of glucose in the body? Where in the body is glycogen stored? When is glycogen used? What does the brain use for energy when glycogen stores are depleted? Under what circumstances are ketone bodies formed? How much dietary carbohydrate is necessary to spare body protein and prevent ketosis? (text, pp. 117–118; video)

9. What is homeostasis? What are the two main regulatory hormones that control blood glucose concentration? What are two conditions that can result when blood glucose regulation fails? What is the "fight-or-flight" response? What is the effect of epinephrine on blood sugar levels? (text, p. 118; video)

10. What are two health effects of excessive sugar intake? What conditions are NOT related to excessive sugar intake? What is the recommended sugar intake based on a percentage of total calories per day? (text, pp. 131–134)

11. What food source should provide most of the energy for active people? What are the long-term and immediate benefits of high carbohydrate diets to athletes? (video)

12. How do carbohydrates affect blood glucose levels? (video)

13. What is the link between sugar intake and hyperactivity and growth? (video)

KEY CONCEPTS

<u>Glucose in the Body</u>

I. Fed State

Glucose —Insulin→

- A. Used for energy (brain)
- B. Stored as glycogen (liver and muscle)
- C. Stored as fat (minor)

II. Fasted State

A. Glycogen	—glucagon→	glucose	(overnight fast)
B. Protein (muscle breakdown)	→	glucose	(1-3 day fast)
C. Fat	→	ketone bodies	(7-10 day fast)

PRACTICE TEST

The following items will help you evaluate your understanding of this lesson. Use the answer key at the end of the lesson to check your answers or to locate material related in each question.

Multiple Choice

Select the one choice that best answers the question.

1. Which of the following is the most important monosaccharide inside the body?
   A. Fructose
   B. Lactose
   C. Glucose
   D. Galactose

2. Which of the following is a polysaccharide?
   A. Starches
   B. White sugar
   C. Lactose
   D. Fructose

3. Which of the following is known as blood sugar?
   A. Glucose
   B. Maltose
   C. Sucrose
   D. Fructose

4. What is the composition of lactose?
   A. Two glucose units
   B. Two fructose units
   C. One glucose and one fructose unit
   D. One glucose and one galactose unit

5. What is the principle carbohydrate of milk?
   A. Lactose
   B. Sucrose
   C. Maltose
   D. Glycogen

6. Glycogen is stored mainly in which of the following tissues?
   A. Muscle and liver
   B. Pancreas and kidneys
   C. Stomach and intestine
   D. Brain and red blood cells

7. The disaccharides include __________.
   A. sucrose, galactose, and maltose
   B. maltose, fructose, and galactose
   C. lactose, glucose, and fructose
   D. sucrose, maltose, and lactose

8. Which of the following enzymes does NOT act on simple sugars?
   A. Lactose
   B. Sucrose
   C. Amylase
   D. Maltose

9. Products of the incomplete breakdown of fat when carbohydrates are not available are called __________.
   A. protein-sparing products
   B. glycogen bodies
   C. ketone bodies
   D. glucagon products

10. What is the current DRI recommendation for carbohydrate intake as a percent of total calories?
   A. Less than 30 percent
   B. 30 to 40 percent
   C. 40 to 50 percent
   D. 45 to 65 percent

11. Athletes or active individuals who require large amounts of energy should consume a diet high in __________.
   A. protein
   B. fat
   C. carbohydrate
   D. water

12. Which of the following statements is the most accurate regarding studies documenting hyperactivity and sugar intake in children?
    A. There is very little evidence to support the statement that sugar causes hyperactivity.
    B. There is considerable evidence that sugar and hyperactivity are positively linked.
    C. There is scientific proof that sugar causes hyperactivity.
    D. None of the above.

13. Hypoglycemia refers to a low level of __________.
    A. glucose
    B. maltose
    C. sucrose
    D. fructose

Matching

Select the letter next to the word or phrase that best matches the numbered statements. Use answers only once.

_____	14. Storage form of glucose in the body	A. Mouth
_____	15. Site where digestion of disaccharides takes place	B. Insulin
		C. Small intestine
_____	16. Substance that signals the release of glucose into blood from the liver	D. Glucagon
_____	17. Site where starch digestion begins	E. Glycogen
_____	18. Hormone that signals the uptake of glucose into cells from the blood	

Essay Questions

Reflect on the items listed below; one or more may be included on the exam.

19. Describe the tasks of the various body systems in breaking down carbohydrate into glucose to fuel the cells' work.

20. List four common accusations of dietary sugar. What is the evidence for and against these accusations.

ANSWER KEY

The following provides the answers for the practice test questions.

	Answer	Learning Objectives	References
1.	C	LO 1, 7	Sizer/Whitney, p. 101
2.	A	LO 2, 3	Sizer/Whitney, p. 102
3.	A	LO 1, 2	Sizer/Whitney, p. 101
4.	D	LO 6	Sizer/Whitney, p. 116
5.	A	LO 2	Sizer/Whitney, p. 101
6.	A	LO 7	Sizer/Whitney, p. 102
7.	D	LO 2	Sizer/Whitney, p. 101
8.	C	LO 2	Sizer/Whitney, p. 115
9.	C	LO 7	Sizer/Whitney, p. 117
10.	D	LO 4	Sizer/Whitney, p. 105
11.	C	LO 7	Video
12.	A	LO 10	Video
13.	A	LO 8	Video
14.	E	LO 5	Sizer/Whitney, p. 102
15.	C	LO 8	Sizer/Whitney, p. 113
16.	D	LO 5	Sizer/Whitney, p. 118
17.	A	LO 8	Sizer/Whitney, p. 113
18.	B	LO 7, 8	Sizer/Whitney, p. 118
19.		LO 1, 7	Sizer/Whitney, p. 118
20.		LO 10	Sizer/Whitney, pp. 131–133

Lesson 4

Carbohydrates: Fiber

OVERVIEW

"You need to eat your roughage," Grandma used to tell us as she put a plate of vegetables in front of us. When we look back on it, we realize that Grandma was very smart! The roughage she was referring to was the indigestible complex carbohydrate, known as fiber, found in vegetables, fruits, grains, and legumes. In this lesson you will not only examine the importance of fiber to the health of the body, but you will also observe what happens to dietary fiber during digestion and absorption. In addition, you will look at the associated side effects of too much fiber in the diet.

In this lesson, you will also delve into the impact of artificial sweeteners in the diet and the different types of natural sweeteners used in the diet and their effects on health.

LEARNING OBJECTIVES

Upon completing this lesson, you should be able to:

1. Compare and contrast the two different types of fiber, including food sources for each type of fiber.
2. Explain how fiber differs from starch.
3. Describe the possible positive health effects of fiber in the diet.
4. Describe the possible adverse effects of fiber in the diet.
5. Cite the recommended intake of fiber.
6. Explain what happens to grain when it is refined into flour.
7. Identify two common sugar substitutes; list two pros and cons for each.

LESSON ASSIGNMENT

Review the following assignment in order to schedule your time appropriately. Pay careful attention; the titles and numbers of the textbook chapter, the telecourse guide lesson, and the video program may be different from one another.

Text: Sizer and Whitney, *Nutrition: Concepts and Controversies*, Chapter 4, "The Carbohydrates: Sugar, Starch, Glycogen, and Fiber," pp. 102–104 and pp. 106–113

Video: "Carbohydrates: Fiber" from the series *Nutrition Pathways*

LESSON FOCUS POINTS

The following questions are designed to help you get the most out of the lesson. After you read the textbook assignments and view the video, you should be able to answer these questions.

1. What is fiber? (text, p. 102)
2. What is the main difference between starch and fiber? How does bacteria affect fiber? Where does most of fiber's action occur in the body? (text, p. 102)
3. What food groups from the Daily Food Guide Pyramid supply fiber? (text, p. 103)
4. How do soluble fibers affect the body? How do insoluble fibers affect the body? What foods provide soluble fiber? What foods provide insoluble fiber? (text, p. 106)
5. How can fiber impact weight control? How do foods rich in soluble fiber affect blood cholesterol and heart disease? How do foods rich in soluble fiber affect diabetes with regard to glucose absorption? Explain how dietary fibers enhance the health of the large intestine with regard to the following: stool size and passage, transit time, constipation, water intake, hemorrhoids, diverticulosis, and colon cancer. (text, pp. 106, 108–109)
6. How can too much fiber affect a person with a small stomach capacity? What symptoms are associated with increasing fiber too quickly? What steps can be taken to prevent complications associated with increased fiber intake? (text, p. 109)

7. What is the recommended intake of fiber per day? What is the average daily fiber intake for people in the United States? (text, p. 109)

8. Define the following terms: refined flour, enriched or fortified flour, and whole grain flour. (text, p. 110)

9. When grain is refined, as in white flour, what part(s) of the grain is (are) removed? Compare grams of fiber in refined and unrefined grains. (text, p. 110; video)

10. What might be the impact of a high-fiber diet on minerals in the body? (text, p. 112)

11. What are the allowed health claims on food labels concerning fiber intake? (text, p. 112)

12. What is a sugar alcohol? What is saccharin? What is aspartame? Do sugar substitutes help with weight control? (text, pp. 134–135, 138)

13. What are the "famous five" conditions that fiber protects us against? How does fiber impact blood cholesterol, risk for heart attack, weight loss, and diabetes? (video)

14. Explain fiber's effects on cancer. (video)

15. How might high dietary fiber intake impact people with Crohn's disease or other inflammatory bowel diseases? (video)

16. What must people do to prevent constipation associated with large fiber intakes? What can happen to people who consume too much fiber or who add fiber too quickly to their diets? (video)

17. What dietary recommendations are encouraged for people with type 2 diabetes regarding fiber and carbohydrate intakes? (video)

A Wheat Plant and a Single Kernel of Wheat

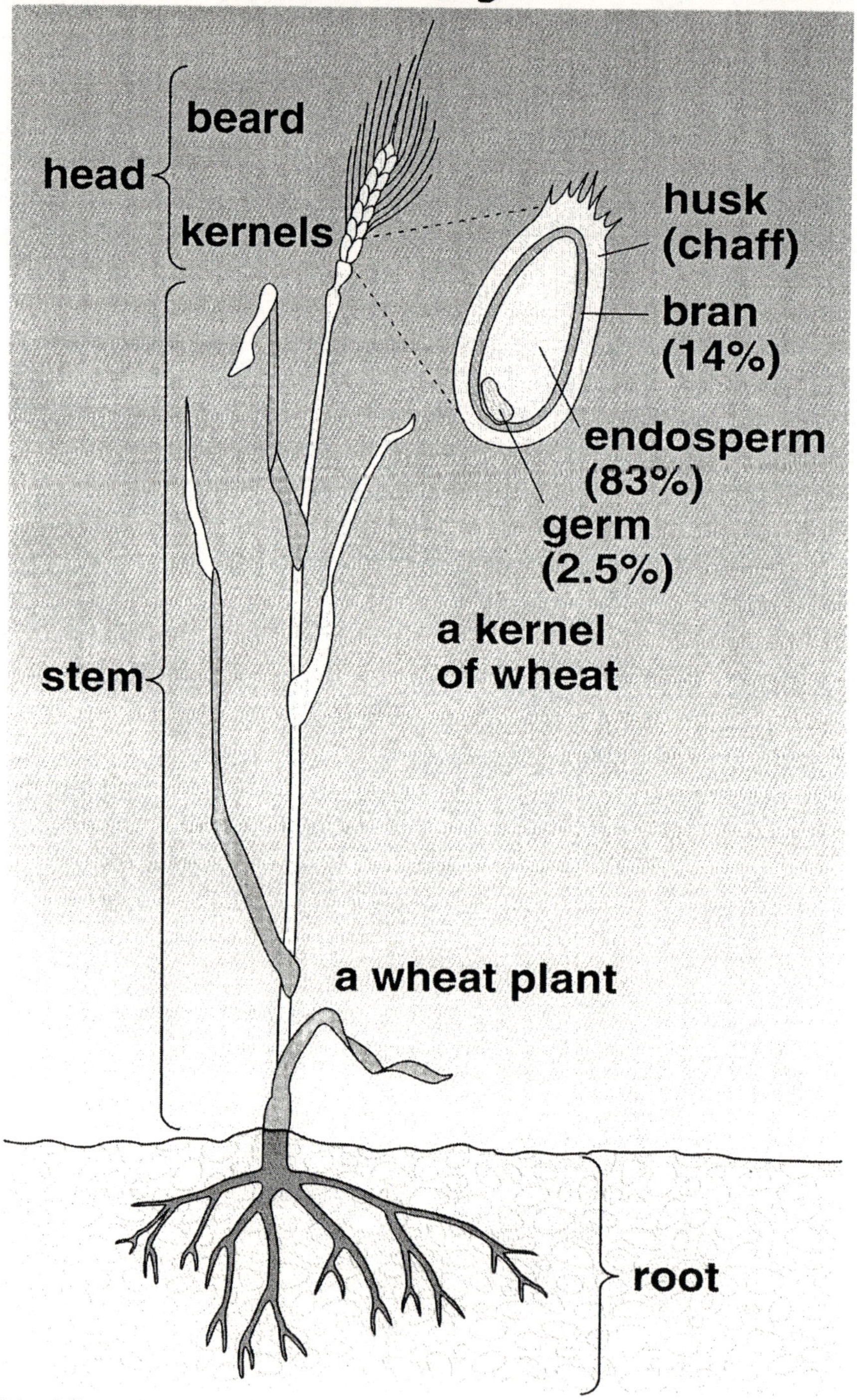

© 2000 Wadsworth/Thomson Learning

Nutrients in Whole-Grain, Enriched White, and Unenriched White Breads

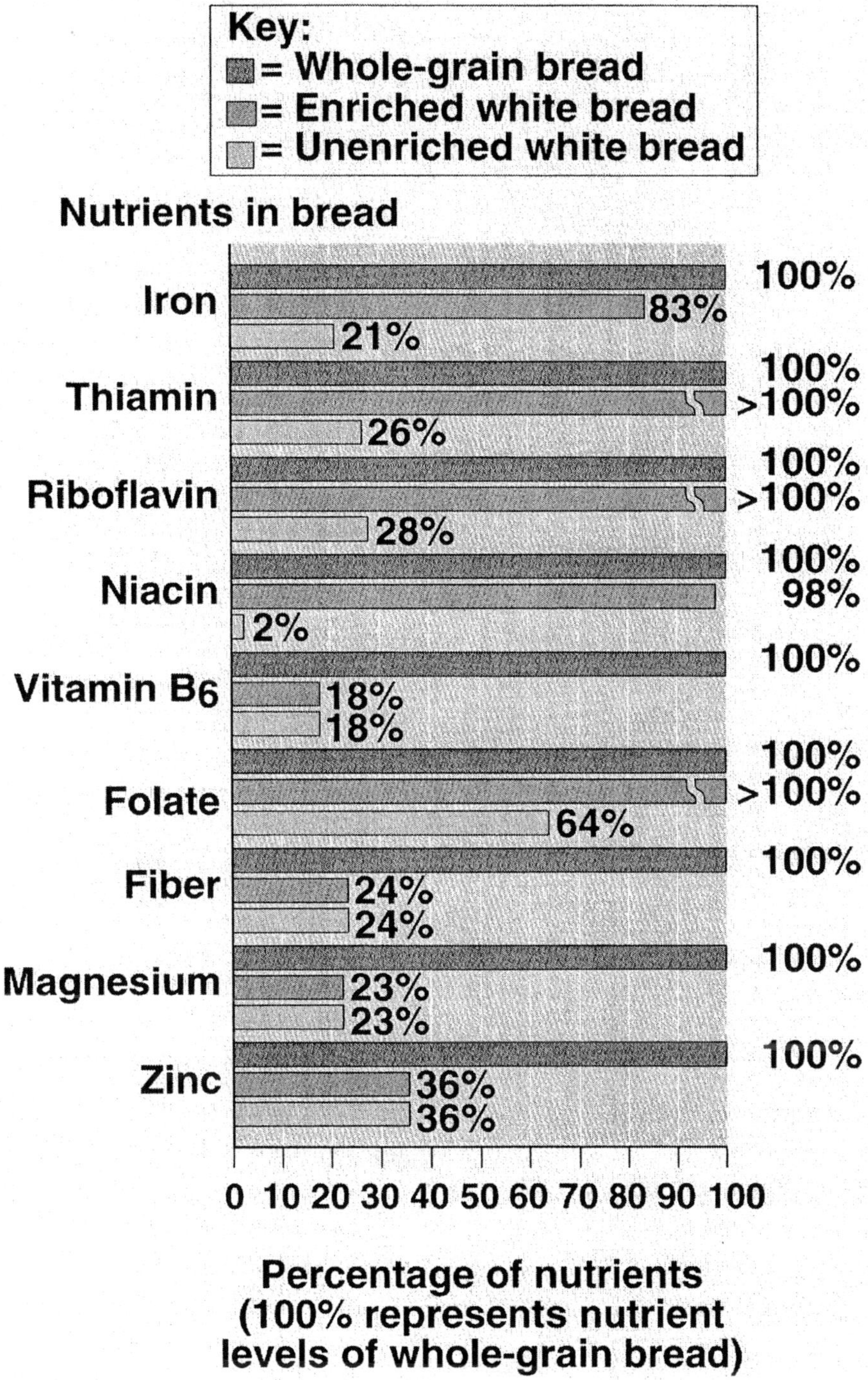

© 2000 Wadsworth/Thomson Learning

Food, Sources, and Health Effects of Fiber

Fiber Type	Major Food Sources	Possible Health Effects
Soluble	▪ Barley ▪ Fruits ▪ Legumes ▪ Oats ▪ Oat Bran ▪ Rye ▪ Seeds ▪ Vegetables	▪ Lower blood cholesterol ▪ Slow glucose absorption ▪ Increase satiety (feel full longer) ▪ Hold moisture in stools, softening them
Insoluble	▪ Brown rice ▪ Fruits ▪ Legumes ▪ Seeds ▪ Vegetables ▪ Wheat bran ▪ Whole grains	▪ Regulate bowel movements by softening stools ▪ Speed transit of material through small intestine ▪ Reduce risks of constipation, diverticulitis, hemorrhoids, and appendicitis ▪ Possibly reduce colon cancer risk

Notes: Excessive fiber intake can decrease absorption of minerals and cause dehydration.

Recommended fiber intake: 25 to 40 grams per day

RELATED ACTIVITY

This activity is not required unless your instructor assigns it. It is offered as a suggestion to help you learn more about the material presented in this lesson.

Complete the form on the following page and return it to your instructor according to the established deadline.

❑ Learning About Fiber in Foods

Learning About Fiber In Food

This exercise is designed to give you practice calculating how much fiber is consumed in a given day. Refer to Chapter 4 of your textbook for help completing this activity.

Directions: Calculate how many grams of fiber are found in each of the following foods, then answer the questions that follow.

Breakfast	Grams of Fiber
1 cup oatmeal	
1 cup skim milk	
1 small apple	
1 tsp sugar	
Lunch	
2 oz lean roast beef	
2 slices white bread	
1 dill pickle	
¾ cup orange juice	
Dinner	
4 oz chicken breast	
½ cup green beans	
1 cup brown rice	
1 cup lowfat frozen yogurt	
Total Fiber	

Cite page numbers in the textbook to support answers to the following questions:

1. According to the American Dietetic Association, what is the daily recommended intake of fiber?

2. Identify the type of fiber that will decrease the risk for CVD and certain cancers.

   CVD:________________________ Cancers:______________________

3. Compare the total fiber found in the foods listed on the previous page to the recommendation stated in Step 1. Was the total fiber content within, below, or above the recommendation?

4. If the total fiber (Step 3) was below the recommendation, name three or more foods and the amount of fiber in each food that would increase the total to the minimum fiber recommendation.

5. Cite two precautions associated with increasing daily fiber intake.

PRACTICE TEST

The following items will help you evaluate your understanding of this lesson. Use the answer key at the end of the lesson to check your answers or to locate material related in each question.

Multiple Choice

Select the one choice that best answers the question.

1. The term *roughage* has been largely replaced by the term __________.
   A. residue
   B. crude fiber
   C. soluble fiber
   D. fiber

2. Which of the following is an example of the difference between the chemical bonds in starch and those in fiber?
   A. Starch bonds are single.
   B. Starch bonds are fatty acids.
   C. Fiber bonds release energy.
   D. Fiber bonds are not broken down during digestion.

3. The fiber-rich portion of the wheat kernel is the bran layer.
   A. True
   B. False

4. Which of the following is NOT an example of fiber?
   A. "Strings" of celery
   B. Membranes surrounding kernels of wheat
   C. Residue in milk
   D. Skins of corn kernels

5. Which of the following foods has the greatest cholesterol-lowering effect?
   A. Wheat bran
   B. Apples
   C. Legumes
   D. Oat bran

6. According to the American Dietetic Association, what is the recommended daily intake of dietary fiber?
   A. 10 to 20 grams
   B. 20 to 35 grams
   C. 55 to 70 grams
   D. 75 to 100 grams

7. Which of the following provides the most fiber?
   A. 1 orange
   B. ½ cup oatmeal
   C. ½ cup lentils
   D. ½ cup green beans

8. Using sugar substitutes has been proven to help people lose weight.
   A. True
   B. False

9. Which of the following foods would you choose as an effective stool-softening agent?
   A. Oat bran
   B. Carrots
   C. Wheat bran
   D. Legumes

10. It would be appropriate to recommended high-fiber foods to someone trying to lose weight because __________.
    A. it speeds up movement of foods through the upper digestive tract
    B. it promotes a feeling of fullness
    C. it may displace calorie-dense concentrated fats
    D. B and C

11. When we eat white bread, the part of the wheat grain that we are eating is the __________.
    A. bran.
    B. germ.
    C. endosperm.
    D. all of the above.

12. High-fiber foods are helpful for people who wish to lose weight because they __________.
    A. take longer to eat
    B. are more filling
    C. are not as well digested
    D. all of the above

13. High levels of dietary fiber intake may protect us from __________.
    A. colon cancer
    B. rectal cancer
    C. breast cancer
    D. all of the above

14. In patients with inflammatory bowel disease or Crohn's disease, high-fiber intakes may benefit them by __________.
    A. normalizing the intestinal contents throughout the colon
    B. causing them to have lower blood pressure which accompanies Crohn's disease
    C. preventing weight gain that accompanies inflammatory bowel disease
    D. all of the above

15. In order to increase fiber intake, people with type 2 diabetes would be encouraged to eat all of the following foods EXCEPT __________.
    A. orange juice
    B. whole oranges
    C. potatoes
    D. broccoli

Fill in the Blank

Insert the correct word or words in the blank for each item.

16. The amount of fiber per serving found in whole wheat kernels and in refined wheat flour, respectively, is _________ and _________.

17. Three of the "famous five" conditions that fiber is known to protect against are ___________________, ___________________, and ___________________.

18. When people increase their dietary fiber intake, they must simultaneously always increase their intake of ________________.

Essay Questions

Reflect on the items listed below; one or more may be included on the exam.

19. List food sources of soluble and insoluble fibers. Contrast the effects of these two types of fibers on gastrointestinal tract function.

20. List two popular sugar substitutes. Discuss the benefits and drawbacks of their use.

ANSWER KEY

The following provides the answers for the practice test questions.

	Answer	Learning Objectives	References
1.	D	LO 1,2	Sizer/Whitney, p. 102
2.	D	LO 2	Sizer/Whitney, p. 103
3.	A	LO 6	Sizer/Whitney, p. 110
4.	C	LO 1	Sizer/Whitney, p. 103
5.	D	LO 1, 3	Sizer/Whitney, p. 106
6.	B	LO 5	Sizer/Whitney, p. 109
7.	C	LO 1	Sizer/Whitney, p. 126
8.	B	LO 7	Sizer/Whitney, p. 138
9.	C	LO 1, 3	Sizer/Whitney, p. 106
10.	D	LO 3	Sizer/Whitney, p. 105
11.	C	LO 6	Video
12.	D	LO 3	Video
13.	D	LO 3	Video
14.	A	LO 3	Video
15.	A	LO 1, 3	Video
16.	9 grams and 1 gram	LO 6	Video
17.	heart disease cancer obesity diabetes high blood pressure	LO 3	Video
18.	water	LO 4	Video
19.		LO 1, 3	Sizer/Whitney, pp. 106–107
20.		LO 7	Sizer/Whitney, pp. 135–137

Lesson 5

Fats: The Lipid Family

OVERVIEW

When people think of fat, they usually think "I've got to reduce the fat in my diet!" or "Fat's bad for you—it causes heart disease!" We typically think of the negative health effects of too much fat in the diet. Fat, however, is not inherently bad. The fact is, without fat there would be very little of the flavor or aroma that we associate with fatty foods and we wouldn't feel nearly as satisfied when we eat if we didn't have some fat. As you make your way through digestion, absorption, and transportation of fat, you will examine the lipoproteins and their functions. You will identify sources of essential fatty acids, including omega-3 fatty acids. Finally, you will examine the origin and function of fat substitutes and how they are used in the diet.

LEARNING OBJECTIVES

Upon completing this lesson, you should be able to:

1. Identify the three main members of the lipid family and identify which of them is the primary lipid in both foods and the body.
2. Explain the ways fats are useful both in the body and in foods.
3. Describe the structure of a triglyceride.
4. Distinguish between saturated, monounsaturated, and polyunsaturated fatty acids, providing examples of food sources for each.
5. Describe the roles of phospholipids and sterols.
6. Summarize the basic steps involved in the digestion, absorption, and transport of lipids.
7. Briefly explain how the body uses stored fat for energy.
8. Identify the functions of cholesterol in the body and indicate why it is not an essential nutrient.

9. Identify the major components and functions of the lipoproteins: chylomicrons, low-density lipoproteins (LDL), and high-density lipoproteins (HDL).
10. Identify the two essential fatty acids, including food sources of omega-3 and omega-6 fatty acids.
11. Describe the process and controversy surrounding hydrogenation.
12. Describe the origin and function of fat substitutes.

LESSON ASSIGNMENT

Review the following assignment in order to schedule your time appropriately. Pay careful attention; the titles and numbers of the textbook chapter, the telecourse guide lesson, and the video program may be different from one another.

Text: Sizer and Whitney, *Nutrition: Concepts and Controversies*, Chapter 5, "The Lipids: Fats, Oils, Phospholipids, and Sterols," pp. 139–178

Video: "Fats: The Lipid Family" from the series *Nutrition Pathways*

LESSON FOCUS POINTS

The following questions are designed to help you get the most out of the lesson. After you read the textbook assignments and view the video, you should be able to answer these questions.

1. What are the three members of the lipid family? Which lipid is the primary one in both foods and in the body? (text, p. 140)
2. What are the functions of triglycerides in the body? What are the functions of triglycerides in foods? (text, pp. 140–142)
3. What is a triglyceride composed of? What is a fatty acid? What is glycerol? (text, p. 142)
4. Define the following fatty acids: saturated, monounsaturated, and polyunsaturated. (text, p. 143)

5. Generally speaking, which lipids are more liquid at room temperature? Why? Which fats are harder at room temperature? Why? (text, p. 143)

6. What is the major fatty acid found in butter? What is the main fatty acid found in olive oil? What is the major fatty acid found in corn oil? (text, p. 145)

7. Why is cholesterol not considered an essential nutrient? What organ manufactures cholesterol? What foods contain cholesterol? What other compounds in the body are made with cholesterol as the starting material? (text, pp. 145–146)

8. Where are triglycerides digested? What is the function of bile? Where is bile stored in the body? What is emulsification? (text, p. 146)

9. What is the general name for the transport vehicles for fats? Describe the main component and function of the following lipoproteins: chylomicron, LDL, and HDL. (text, pp. 147–149)

10. When is stored fat used for energy? Why are carbohydrates necessary in using fat for energy? What is produced when carbohydrates are not available when using fat for energy? How does the body store excess calories as fat? Where is excess fat stored? (text, pp. 147–149)

11. Why are lipids transported in lipoproteins? (text, p. 149; video)

12. What are the two essential fatty acids that the body cannot make? Which one is omega-6? Which one is omega-3? What foods provide omega-6 fatty acids in the diet? What are good food sources of omega-3 fatty acids? (text, pp. 154–156)

13. What is hydrogenation? What is a *trans* fatty acid? What are the advantages and disadvantages of hydrogenation? (text, pp. 158–159)

14. Why were fat substitutes developed? Define Simplesse and olestra. (text, p. 162; video)

15. Why are triglycerides considered the best energy source for strenuous activities, such as hiking? What are examples of high-fat foods that would support an activity such as hiking? (video)

16. When does cholesterol become harmful to the body? In order to deal with problems associated with cholesterol, what two components must be considered?

What dietary component can significantly increase LDL, or the "bad" cholesterol? What lifestyle changes can improve blood cholesterol? (video)

17. What is the recommended limit of blood cholesterol? What dietary factors can increase triglyceride levels? What happens to an artery that has cholesterol deposits? What dietary components must be monitored for someone who has high blood cholesterol? (video)

KEY CONCEPTS

Fats in Food

I. Nutrition
- Fats provide essential fatty acids.
- Fats provide a concentrated energy source in foods.
- Fats carry fat-soluble vitamins A, D, E, and K, and assist in their absorption.
- Fats provide raw material for making needed products.

II. Taste
- Fats contribute to the taste and smell of foods.
- Fats stimulate the appetite.
- Fats contribute to feelings of fullness (also called satiety).
- Fats help make foods tender.

Fats in the Body

- *Energy stores.* Fats are the body's chief form of stored energy.
- *Muscle fuel.* Fats provide most of the energy to fuel muscular work.
- *Emergency reserve.* Fats serve as an emergency fuel supply in times of illness and diminished food intake.
- *Padding.* Fats protect the internal organs from shock through fat pads inside the body cavity.
- *Insulation.* Fats insulate against temperature extremes through a fat layer under the skin.
- *Cell membranes.* Fats form the major material of cell membranes.
- *Raw materials.* Fats are converted to other compounds, such as hormones, bile, and vitamin D, as needed.

A Closer Look at Food Fats

I. Triglyceride Structure

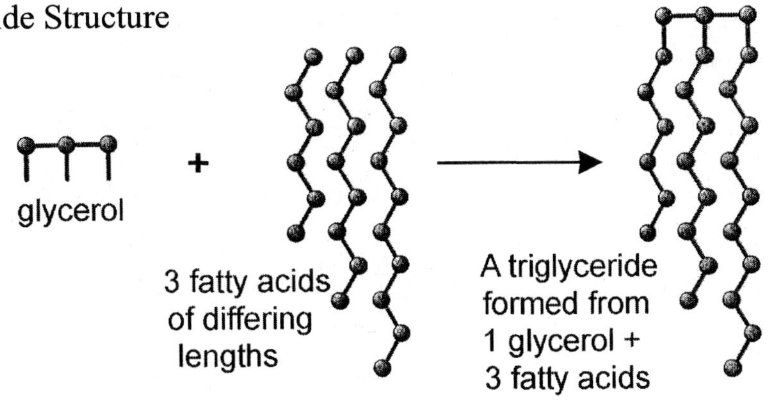

- Contains one glycerol and three fatty acids.
- Fatty acids may be saturated or unsaturated.
- *Un*saturated fatty acids may be *mono*unsaturated or *poly*unsaturated.
- Fats in foods are classified according to whichever fatty acid is present in the largest amount, even though food fats contains some of all three fatty acids.

II. Common Food Sources of Fat

The more carbon atoms in a fatty acid, the longer it is. The more hydrogen atoms attached to those carbons, the more saturated the fatty acid is.

A. Saturated
- Palm oil
- Coconut oil
- Butter

B. Monounsaturated
- Olive oil
- Peanut oil
- Canola oil

C. Polyunsaturated
- Safflower oil
- Sunflower oil
- Corn oil
- Soybean oil
- Cottonseed oil

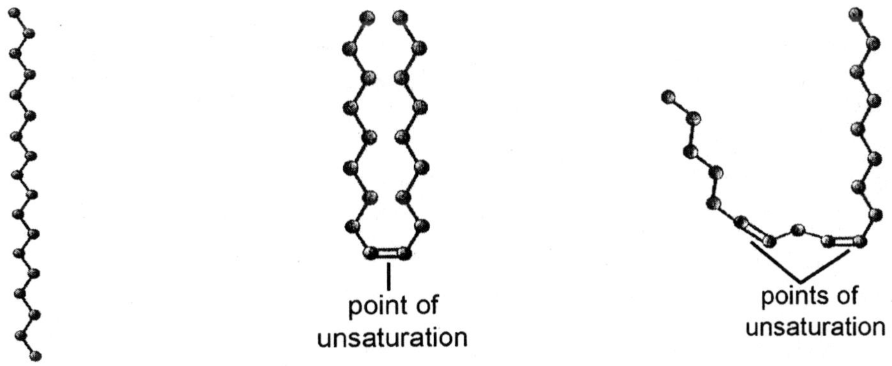

III. Hydrogenation

Example: Hydrogens are artificially added to a vegetable oil.

Polyunsaturated fatty acid (soybean oil) —Hydrogen (H+)→ *trans* fatty acid (partially hydrogenated soybean oil)

Note that an unsaturated *trans* fatty acid is similar in shape to a saturated fatty acid; it behaves similarly in the body too.

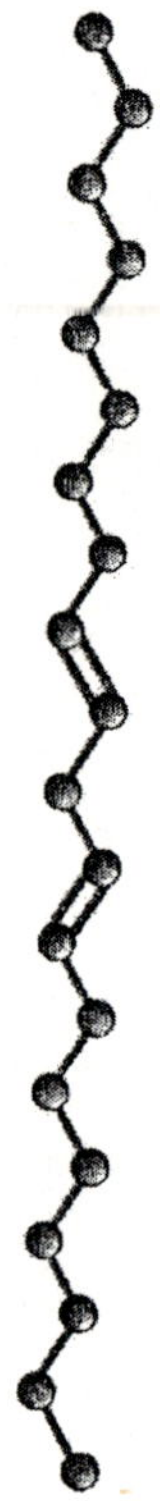

Results:

- Fat is more solid.
- Fat is spreadable.
- Fat has longer shelf life.
- Fat is bad for heart disease.

Lipid Digestion, Absorption, and Transport

The double bonds that remain in the fatty acid have a different shape than the ones naturally present in the oil, and are called *trans* fatty acids. This type of fat is not usually found in nature.

I. Lipid Digestion
   Triglyceride (bile, lipase) ⟶ fatty acids + glycerol (inside small intestine)

II. Lipid Absorption
   A. Fatty acids and glycerol are absorbed into intestinal cells.
   B. Fatty acids + glycerol ⟶ triglyceride (inside intestinal cells)
   C. Triglycerides are packed together into chylomicrons.
   D. Chylomicrons are released into the lymph vessels and then into the bloodstream.

III. Lipoproteins
   A. Chylomicron
      - Formed from dietary fat
      - Contains 80 percent triglycerides
   B. Very-low-density lipoproteins (VLDL)
      - Produced by liver
      - Contains 50 percent triglycerides
   C. Low-density lipoproteins (LDL)
      - Also known as "bad" cholesterol
      - Formed from VLDL when triglycerides are removed by cells
      - Contains 50 percent cholesterol
   D. High-density lipoproteins (HDL)
      - Also known as "good" cholcstcrol
      - Produced by the liver
      - Transports cholesterol from cells to liver to be excreted
      - Contains 50 percent protein

IV. Lipid Metabolism

A. Using fat for energy

- With adequate glucose
  Fat ⟶ Energy (ATP)

- Without glucose (fasted)
  Fat ⟶ ketones

B. Storing fat to turn either carbohydrates or proteins into triglycerides and store as body fat requires energy.

- Fat in diet (triglycerides) ⟶ body fat (triglycerides)
- Excess carbohydrates (glucose) ⟶ body fat (triglycerides)
- Excess protein (amino acids) ⟶ body fat (triglycerides)

PRACTICE TEST

The following items will help you evaluate your understanding of this lesson. Use the answer key at the end of the lesson to check your answers or to locate material related in each question.

Multiple Choice

Select the one choice that best answers the question.

1. Which type of fatty acid is found in high amounts in olive oil?
   A. Saturated
   B. Monounsaturated
   C. Polyunsaturated
   D. Partially hydrogenated

2. Characteristics of hydrogenated oils include all of the following EXCEPT ___.
   A. they are stored in adipose tissue.
   B. they lower HDL and raise LDL cholesterol in the body.
   C. they contain *trans* fatty acids.
   D. products containing them become rancid sooner, contributing to a shorter shelf life.

3. Which of the following is the best source of omega-3 fatty acids?
   A. Tuna
   B. Butter
   C. Salad oil
   D. Shortening

4. What lipoprotein acts as a scavenger and is responsible for transporting cholesterol back to thc liver?
   A. Chylomicron
   B. Low-density lipoprotein
   C. High-density lipoprotein
   D. Very-low-density lipoprotein

5. Characteristics of the essential fatty acids include __________.
   A. they must be supplied by the diet
   B. they can be made from the substances in the body
   C. they are polyunsaturated fatty acids
   D. A and C

6. About 95 percent of the lipids in foods and in the human body are __________.
   A. triglycerides
   B. phospholipids
   C. sterols
   D. cholesterol

7. Which of the following is NOT a function of fat in the human body?
   A. Protects the body from temperature extremes
   B. Cushions the internal organs from physical shock
   C. Carries the water-soluble nutrients
   D. Provides the major material from which cell membranes are made

8. Which of the following statements about *trans* fatty acids is TRUE?
   A. They are made by the body.
   B. Their amounts are listed on foods labels.
   C. They arise when polyunsaturated oils are hydrogenated.
   D. They occur naturally in foods to a large extent.

9. Characteristics of the artificial fat olestra include __________.
   A. it passes through the digestive tract unabsorbed
   B. it is fortified with vitamin E
   C. it has undesirable side effects
   D. all of the above

10. The best energy source for strenuous, aerobic type activities such as hiking is __________.
   A. phospholipids
   B. triglycerides
   C. carbohydrates
   D. sterols

11. Lipoproteins __________.
   A. transport lipids through a watery medium, the blood
   B. are protein-based vehicles for the transport of lipids
   C. are composed of lipids and proteins
   D. all of the above

12. Blood cholesterol is impacted by __________.
   A. lifestyle
   B. nutrition
   C. heredity
   D. all of the above

13. Fat substitutes were developed to __________.
   A. help control calories in foods
   B. help people reduce body fat
   C. make foods taste better
   D. all of the above

14. Triglyceride levels can be reduced by reducing which of the following foods?
    A. Alcohol
    B. Sweets
    C. Complex carbohydrates
    D. A and B

Matching

Select the letter next to the word or phrase that best matches the numbered statements. Use answers only once.

_____	15. A lipid that is solid at room temperature	A. Cholesterol
_____	16. Contains three fatty acids and one glycerol	B. Olestra
_____	17. One of the sterols that is used to make vitamin D	C. Animal fat
		D. Triglyceride
_____	18. A lipid that is liquid at room temperature	E. Vegetable oil

Essay Questions

Reflect on the items listed below; one or more may be included on the exam.

19. Describe the process of fat hydrogenation and discuss its advantages and disadvantages.

20. Discuss in detail the digestion, absorption, and transport of dietary fat.

ANSWER KEY

The following provides the answers and references for the practice test questions.

	Answer	Learning Objectives	References
1.	B	LO 4	Sizer/Whitney, p. 143
2.	D	LO 11	Sizer/Whitney, p. 159
3.	A	LO 10	Sizer/Whitney, p. 155
4.	C	LO 9	Sizer/Whitney, p. 149
5.	D	LO 10	Sizer/Whitney, p. 155
6.	A	LO 1	Sizer/Whitney, p. 140
7.	C	LO 2	Sizer/Whitney, p. 140
8.	C	LO 11	Sizer/Whitney, p. 159
9.	D	LO 12	Sizer/Whitney, pp. 162–163
10.	B	LO 7	Video
11.	D	LO 6, 9	Video
12.	D	LO 9	Video
13.	A	LO 12	Video
14.	D	LO 9	Video
15.	C	LO 4	Sizer/Whitney, p. 143
16.	D	LO 1, 3	Sizer/Whitney, p. 142
17.	A	LO 5	Sizer/Whitney, p. 145
18.	E	LO 4	Sizer/Whitney, p. 143
19.		LO 11	Sizer/Whitney, p. 158
20.		LO 1, 6	Sizer/Whitney, pp. 146–148

Lesson 6

Fats: Health Effects

OVERVIEW

The role of fat in the diet is important for the normal functioning of the body. What are the consequences of too much fat in the diet? High fat intake has been linked to cardiovascular disease, diabetes, obesity, and many cancers. This lesson will examine the health effects of fatty acid saturation and high dietary cholesterol intake. Further, it will examine how blood lipid profiles (which reveal LDL, triglycerides, total serum cholesterol, and HDL levels) are used to detect risks for diseases. Finally, it will look at strategies to improve the health risks associated with high fat intakes.

LEARNING OBJECTIVES

Upon completing this lesson, you should be able to:

1. Describe the purpose of a blood lipid profile.
2. Describe the health effects of exercise and dietary fats on blood.
3. Cite the recommended intakes of total dietary fat, saturated fat, and dietary cholesterol.
4. Suggest practical ways to reduce total dietary fat, saturated fat, and dietary cholesterol.
5. Given specific high-fat food choices, provide examples of lower-fat alternatives to high-fat foods.
6. Suggest ways to balance omega-3 and omega-6 fatty acid intakes.
7. Differentiate between the health effects of a traditional meat-based diet of the United States and a traditional olive oil-based diet of the Mediterranean.
8. Read food labels to determine type of fat and fat content of packaged foods.
9. Calculate the percentage of calories and grams from daily fat intakes.

LESSON ASSIGNMENT

Review the following assignment in order to schedule your time appropriately. Pay careful attention; the titles and numbers of the textbook chapter, the telecourse guide lesson, and the video program may be different from one another.

Text: Sizer and Whitney, *Nutrition: Concepts and Controversies*, Chapter 5, "The Lipids: Fats, Oils, Phospholipids, and Sterols," pp. 139–178; Appendix C, D

Video: "Fats: Health Effects" from the series *Nutrition Pathways*

LESSON FOCUS POINTS

The following questions are designed to help you get the most out of the lesson. After you read the textbook assignments and view the video, you should be able to answer these questions.

1. What is the purpose of a blood lipid profile? (text, p. 140)
2. What are the daily amounts both the Committee on Diet and Health and the proposed Healthy People 2010 recommend for the following: percentage of total fat intake, saturated fat intake, and dietary cholesterol intake? (text, p. 150)
3. What is the percentage of calories from fat recommended by the American Heart Association? What is considered the best type of fat to select? What is the maximum percent of total fat that should be in the form of saturated fat? (text, p. 150; video)
4. Name the dietary lipid that raises blood cholesterol even more dramatically than dietary cholesterol does. What type of fat generally raises LDL? What are the risks associated with *trans* fatty acid intakes? (text, pp. 150–152, Appendix C)
5. What calculations are necessary to determine the percentage of fat in foods? What calculations are necessary to determine grams of fat in the diet? What calculations are necessary to determine the percentage of fat in the diet? (text, pp. 153, 166–167, Appendix C; video)
6. What strategy may be more effective in preventing heart disease even more than reducing total fat intake? In general, how do polyunsaturated and

monounsaturated fatty acids affect LDL and HDL levels? What does research reveal about fish intake with regard to risk for cardiovascular disease (CVD)? (text, pp. 154–157)

7. What is the current advice for balancing dietary intakes of omega-3 and omega-6 fatty acids? Why are fish oil supplements not recommended? (text, pp. 155–157)

8. What information regarding fat content is omitted from food labels? (text, p. 160)

9. What are some steps a person can take to reduce total dietary fat, saturated fat, and dietary cholesterol? What are "added fats"? Which two groups in the Food Pyramid always contain fat? Which two groups sometimes contain fat? Which two groups rarely contain fat? What are five principles to follow when trying to decrease dietary fat? (text, pp. 160–161; video)

10. Name at least one low-fat alternative for each of the following high-fat foods: whole milk, sour cream, cheese, salad dressing, and bacon. (text, pp. 165–169; appendix D)

11. What is the basis of a Mediterranean diet? What health-related benefits may occur in people who consume a Mediterranean diet? How can olive oil and fish play a role in a healthy diet? (text, pp. 174–177; video)

12. What are the differences between the USDA Food Guide Pyramid and the Mediterranean Diet Pyramid? What are the similarities? (text, p. 177)

13. What steps can be taken to reduce fat in meat-based diets? What health-related problems can occur in people who eat high-meat/fat diets? How can a higher-fat diet be justified? (video)

14. Explain how to reduce fat from the following foods: 8 ounces of prime rib, mashed potatoes with milk and butter, fried zucchini, garlic bread with butter, a salad with Italian dressing, and a piece of apple pie. How can reducing fat in the diet affect one's overall health? (appendix D, video)

15. What steps can be taken by a person with type 2 diabetes to reduce fat intake? (video)

KEY CONCEPTS

Recommended Intakes for Fat

Types of Fat	Recommendation
Total fat	Keep between 20–35%
Saturated fat	Less than 10% of calories
Cholesterol	Less than 300 mg per day
Omega-3 and omega-6	Balance by eating fish two to three times per week. (No fish oil supplements)

Effect of Dietary Fat and Exercise on Blood Lipids

Component	Effect
Saturated fat	Raises LDL
Monounsaturated fat	Lowers LDL
Polyunsaturated fat: omega-6	Lowers LDL
Polyunsaturated fat: omega-3	Decreases blood clotting
Trans fat: partially hydrogenated	Raises LDL; lowers HDL
Cholesterol in foods	Varies; not as important as amount and type of fat
Exercise	Raises HDL

RELATED ACTIVITY

This activity is not required unless your instructor assigns it. It is offered as a suggestion to help you learn more about the material presented in this lesson.

Complete the form on the following page and return it to your instructor according to the established deadline.

- ❑ Determining Percentage of Calories and Fat Grams from Fat Intakes

Determining Percentage of Calories and Fat Grams from Fat Intakes

The purpose of this exercise is to give you practice calculating grams of fat, fat calories, and percentage of total calories in specific foods. Refer to Chapter 5 of your textbook for help completing this activity.

PROBLEM #1: A three-ounce pork chop has a total of 275 calories with 8 fat grams per ounce.

1. ____________ How many calories are in 1 gram of fat?

2. ____________ Multiply fat grams per ounce by the number of ounces of pork, then multiply by number of calories in 1 gram of fat to find how many total fat calories are in the pork chop. (Show your calculations below.)

3. ____________ Divide total fat calories (step #2) by total calories in the pork chop to find the percentage of total calories from fat. (Show your calculations below.)

PROBLEM #2: A man consumes 2,500 total calories per day and 110 grams of fat.

4. ____________ Multiply grams of fat by number of calories in 1 gram of fat, then divide by the total calories per day to find the percentage of total calories from fat the man consumes. (Show your calculations below.)

PROBLEM #3: A woman consumes 2,200 total calories per day and 30 percent of the calories come from fat.

5. ____________ Multiply the total calories per day by 0.30, then divide by number of calories in 1 gram of fat to find the number of grams of fat the woman consumes. (Show your calculations below.)

PRACTICE TEST

The following items will help you evaluate your understanding of this lesson. Use the answer key at the end of the lesson to check your answers or to locate material related in each question.

Multiple Choice

Select the one choice that best answers the question.

1. A low risk of cardiovascular disease correlates with high blood levels of __________.
   A. triglycerides
   B. free fatty acids
   C. high-density lipoproteins
   D. very-low-density lipoproteins

2. How many grams of fat and total calories from fat are contained in this meal: one scrambled egg made with 1 teaspoon margarine; two strips of bacon; one slice of whole wheat toast with 1 teaspoon margarine; half a grapefruit; and one cup of 2 percent milk?
   A. 1,000
   B. 2,500
   C. 3,500
   D. 4,000

3. The best way to increase consumption of omega-3 fatty acids is to __________.
   A. increase intake of seed oils
   B. consume two to three fish meals per week
   C. limit intake of polyunsaturated fats
   D. take fish oil in supplements

4. You are talking with someone who is trying to decrease his intake of saturated fat. Which of the following items would you recommend as a coffee whitener?
   A. cream
   B. non-dairy creamer
   C. skim milk
   D. half and half

5. The main purpose of the blood lipid profile is to __________.
   A. determine a person's risk for developing diabetes
   B. reveal a person's risk of CVD
   C. determine a person's genetic predisposition for developing kidney disease
   D. reveal how much food cholesterol is in a person's blood

6. Approximately what percent calories from fat would be provided by a soup with 200 calories per serving and 99 calories from fat?
   A. 44 percent
   B. 50 percent
   C. 67 percent
   D. 72 percent

7. According to the American Heart Association, what is the upper limit of fat that should be consumed by a healthy person requiring 3,000 calories per day?
   A. 33 grams
   B. 50 grams
   C. 60 grams
   D. 100 grams

8. According to the American Heart Association, what should be the maximum daily intake of cholesterol on a 2,500 calorie diet?
   A. 50 milligrams
   B. 150 milligrams
   C. 300 milligrams
   D. 1,000 milligrams

9. In an effort to lower LDL, you would tell someone to __________.
   A. use olive oil in the place of saturated fat
   B. use vegetable shortening in the place of lard
   C. use corn oil in the place of vegetable shortening
   D. A and C

10. The American Heart Association recommends the percentage of total calories that come from fat should not exceed __________.
    A. 10 percent
    B. 20 percent
    C. 30 percent
    D. 40 percent

11. A person can reduce fat in a diet that has meat as its base by __________.
   A. selecting leaner cuts of meat
   B. choosing chicken or fish occasionally
   C. increasing the intake of vegetable oils
   D. both A and B

12. Some people who eat high-meat diets may have a greater risk for __________.
   A. cardiovascular disease
   B. colon cancer
   C. obesity
   D. all of the above

13. Higher-fat intakes can be justified if a person is __________.
   A. doing aerobics for short periods of time
   B. over the age of fifty-five
   C. performing manual labor for long periods of time
   D. under the age of fifty-five

14. The basis of a Mediterranean diet consists of __________.
   A. olive oil as the primary fat
   B. fruits and vegetables as the basis
   C. lots of grains and breads
   D. all of the above

15. Most people who consume a Mediterranean diet usually __________.
   A. suffer from obesity
   B. do not suffer from cardiovascular disease
   C. show signs of type II diabetes
   D. do not develop lung cancer

16. If you wanted to reduce the fat and calories in a 10-ounce portion of prime rib, you could __________.
   A. eat 5 ounces and take the rest home for the next day
   B. cut away all visible fat but still eat the 10 ounces
   C. refuse to eat any of the meat
   D. either A or B

17. Reducing fat intake for a person with type 2 diabetes can be accomplished by __________.
    A. increasing fruits and vegetables
    B. increasing meat portions
    C. decreasing breads and grains
    D. none of the above

Essay Questions

Reflect on the items listed below; one or more may be included on the exam.

18. Discuss the role of dietary cholesterol and the production of HDL and LDL cholesterol on heart disease. What is meant by "good" and "bad" cholesterol?

19. List strategies for lowering fat intake and still allowing for eating enjoyment.

ANSWER KEY

The following provides the answers and references for the practice test questions.

Answer	Learning Objectives	References
1. C	LO 2	Sizer/Whitney, p. 149
2. C	LO 9	Sizer/Whitney, Appendix C
3. B	LO 6	Sizer/Whitney, p. 155
4. C	LO 4, 5	Sizer/Whitney, p. 161
5. B	LO 1	Sizer/Whitney, p. 140
6. B	LO 8	Sizer/Whitney, pp. 160–162
7. D	LO 3, 9	Sizer/Whitney, p. 150
8. C	LO 3	Sizer/Whitney, p. 150
9. D	LO 2	Sizer/Whitney, p. 176
10. C	LO 3	Sizer/Whitney, p. 150; Video
11. D	LO 4, 5	Video
12. D	LO 2	Video
13. C	LO 2, 7	Video
14. D	LO 7	Video
15. B	LO 7	Video
16. A	LO 4, 5	Video
17. A	LO 4	Video
18.	LO 1, 2	Sizer/Whitney, p. 150
19.	LO 4, 5	Sizer/Whitney, pp. 168–171

Lesson 7

Protein: Form and Function

OVERVIEW

If you were to ask people, "What is the function of protein in the body?" most would undoubtedly say that protein is used to build strong muscles. The truth is, however, that building muscle is just one of the many varied jobs that protein does in the body, as you will see in this lesson. You will discover why some amino acids are essential nutrients As you follow food protein through the GI tract, you will observe what happens when protein is digested, absorbed, and metabolized. Finally, you will see how protein-energy malnutrition (PEM) affects the health and growth of those who suffer from protein-energy malnutrition.

LEARNING OBJECTIVES

Upon completing this lesson, you should be able to:

1. Identify the element that is unique to protein.
2. Define essential amino acid.
3. Describe the basic digestion and absorption of protein.
4. List the basic steps in protein synthesis.
5. Explain what happens when sequencing errors occur during protein synthesis.
6. Describe the roles of protein in the body.
7. Describe the various fates of an amino acid inside the cell.
8. Distinguish between nitrogen equilibrium, positive nitrogen balance, and negative nitrogen balance.
9. Describe the characteristics of protein-energy malnutrition (PEM).
10. Cite possible health effects of excessive protein intake.

LESSON ASSIGNMENT

Review the following assignment in order to schedule your time appropriately. Pay careful attention; the titles and numbers of the textbook chapter, the telecourse guide lesson, and the video program may be different from one another.

Text: Sizer and Whitney, *Nutrition: Concepts and Controversies*, Chapter 6, "The Proteins and Amino Acids," pp. 179–212

Video: "Protein: Form and Function" from the series *Nutrition Pathways*

LESSON FOCUS POINTS

The following questions are designed to help you get the most out of the lesson. After you read the textbook assignments and view the video, you should be able to answer these questions.

1. What atoms are found in all proteins, and which one of these atoms is not found in either carbohydrate or fat? What distinguishes each amino acid from other amino acids? (text, p. 180)

2. How many amino acids make up proteins? What makes an amino acid "essential"? How many amino acids are essential? (text, p. 180)

3. What is the function of deoxyribonucleic acid (DNA) in the synthesis of new proteins? What is the function of the messenger ribonucleic acid (RNA)? What is the function of transfer RNA? What is a peptide bond? (text, pp. 181–186)

4. What can happen if a protein sequence is altered? What is the disease associated with a sequencing error in hemoglobin? (text, p. 183)

5. Where does the digestion of protein begin? What happens to protein in the stomach? What is the function of hydrochloric acid? What is pepsin? What happens to polypeptides when they enter the small intestine? Where does absorption of amino acids take place? (text, pp. 186–187)

6. Summarize the functions of protein. (text, pp. 188–192)

7. What are the various ways a cell can use an amino acid? Under what circumstance will amino acids be used for energy? What is deamination? What happens if a person eats more protein than the body needs? How is nitrogen excreted? What is urea? (text, pp. 191–192)

8. What is nitrogen balance? What is nitrogen equilibrium? Provide examples of people in positive nitrogen balance. Provide examples of people in negative nitrogen balance. Is there a storage form of protein in the body? (text, pp. 197–198)

9. What is protein-energy malnutrition (PEM)? How many children worldwide are affected by PEM? Where is PEM most prevalent in the world? How does PEM affect the United States? What is marasmus? What population is most affected by marasmus? What are the symptoms of marasmus? What is kwashiorkor? What population is most affected by kwashiorkor? What are the symptoms associated with kwashiorkor? How should nutrition therapy be applied in children with PEM? (text, pp. 198–201)

10. What is a typical diet of children with PEM in third world countries? How does PEM impact health in third world countries and in the United States? (text, pp. 198–201; video)

11. How does excessive protein intake impact health? (text, p. 201; video)

12. What conditions and diseases have been correlated with high-protein intake? How does a high-protein diet contribute to heart disease? (text, pp. 201–202)

13. Cite inaccurate statements regarding the function of protein. What is a sound diet for bodybuilders? (video)

14. How can federally funded programs in the United States help prevent PEM in school children? (video)

15. What can people do regarding protein intake if they have high blood cholesterol? (text, p. 208; video)

KEY CONCEPTS

<u>Digestion of Protein</u>

Dietary Protein $\xrightarrow[\text{stomach}]{\text{HCl}}$ Denatured protein (uncoiled)

Denatured protein $\xrightarrow[\text{stomach}]{\text{protease}}$ Polypeptides

Polypeptides $\xrightarrow[\text{small intestine}]{\text{proteases}}$ Amino acids

<u>Amino Acids in the Body</u>

I. Adequate Energy from Carbohydrates and Fat

Amino acids → A. Make other amino acids

Amino acids → B. Make protein

Amino acids → C. Make other compounds

II. Inadequate Energy from Carbohydrates and Fat
   A. Remove nitrogen
   B. Amino acids (from diet or body) → glucose

III. Excess Protein Intake
   A. Remove nitrogen
   B. Amino acids $\xrightarrow{\text{uses energy}}$ glycogen or fat

IV. What Happens to Nitrogen Removed from Amino Acids?

Nitrogen from protein —(liver)→ urea —(kidney)→ excreted in urine

V. Steps in Protein Synthesis

A. DNA inside the nucleus makes an exact copy of itself called messenger RNA.
B. Messenger RNA leaves nucleus and goes to the ribosome inside the cell. The ribosome is where the new protein is made.
C. Many transfer RNAs pick up amino acids from the cell fluid.
D. The transfer RNAs take all these varied amino acids to the ribosome.
E. The messenger RNA then directs the order of the amino acids as they are linked together. (Remember there are twenty different amino acids, and each protein requires them to be in exactly the right order. There are twenty-six letters in the English alphabet and think how many words there are!)
F. A new protein with a specific function inside the body is made. Some examples are enzymes, antibodies, hormones, etc.

The Importance of Amino Acids

Just as each letter of the alphabet is important in forming whole words, each amino acid must be available to build finished proteins.

What happens if the amino acids are not linked in the proper order?

If just one amino acid is out of order, then the protein cannot function properly.

How does this happen?

If there is a problem with the DNA, then the problem is passed to the messenger RNA since it is an exact copy of the DNA. There could also be a problem when the messenger RNA is copied. This causes an error in the order of the amino acids as they are linked together.

Can you give an example?

Sickle cell disease is one example. Normally red blood cells are round. In sickle cell disease the red blood cell has a crescent shape, which interferes with transport of oxygen and blood flow. All this because of one slight alteration in the sequence of amino acids in making the protein hemoglobin.

Normal Red Blood Cells and Sickle Cells

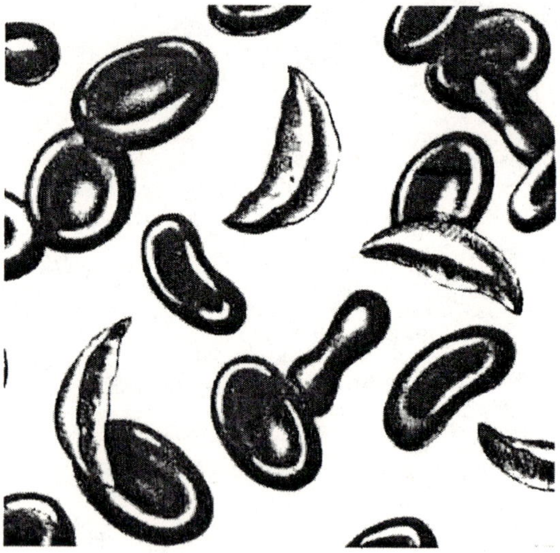

Summary of Proteins' Functions

Growth and maintenance	**Proteins form integral parts of skin, tendons, membranes, muscles, organs, and bones. As such, they support the growth and repair of body tissues.**
Enzymes	**Proteins facilitate chemical reactions.**
Hormones	**Proteins regulate body processes. (Some hormones are made of protein.)**
Antibodies	**Proteins inactivate foreign invaders and protect the body against diseases.**
Fluid and electrolyte balance	**Proteins help maintain the volume and composition of body fluids.**
Acid-base balance	**Proteins help maintain the acid-base balance of body fluids by acting as buffers.**
Transportation	**Proteins transport substances, such as lipids, vitamins, minerals, and oxygen, around the body.**
Energy	**Proteins provide some fuel for the body's energy needs.**

© 1999 Wadsworth Publishing Company/ITP

Features of Marasmus and Kwashiorkor in Children

Separating PEM into two classifications oversimplifies the condition, but at the extremes, marasmus and kwashiorkor exhibit marked differences. Marasmus-kwashiorkor mix presents symptoms common to both marasmus and kwashiorkor. In all cases, children are likely to develop diarrhea, infections, and multiple nutrient deficiencies.

Marasmus

- Infancy (less than two years old)
- Severe deprivation or impaired absorption of protein energy, vitamins, and minerals
- Develops slowly; chronic PEM
- Severe weight loss
- Severe muscle wasting with fat loss
- Growth is less than 60 percent weight for age
- No detectable edema
- No fatty liver
- Anxiety, apathy
- Appetite may be normal or impaired
- Hair is sparse, thin, and dry, easily pulled out
- Skin is dry, thin, and wrinkled

Kwashiorkor

- Older infants and younger children (one to three years old)
- Inadequate protein intake or, more commonly, infections
- Rapid onset; acute PEM
- Some weight loss
- Some muscle wasting, with retention of some body fat
- Growth: 60 to 80 percent weight for age
- Edema
- Enlarged, fatty liver
- Apathy, misery, irritability, sadness
- Loss of appetite
- Hair is dry and brittle, easily pulled out; changes color; becomes straight
- Skin develops lesions

PRACTICE TEST

The following items will help you evaluate your understanding of this lesson. Use the answer key at the end of the lesson to check your answers or to locate material related in each question.

Multiple Choice

Select the one choice that best answers the question.

1. What element is found in proteins but NOT in carbohydrates and fats?
   A. Carbon
   B. Oxygen
   C. Calcium
   D. Nitrogen

2. How many amino acids are considered to be essential amino acids?
   A. Five
   B. Seven
   C. Nine
   D. Thirteen

3. A(n) ______ bond is formed between the amino end of one amino acid and the acid end of the next amino acid in a protein.
   A. peptide
   B. amino acid
   C. denatured
   D. sulfur

4. The sequences of amino acids that make up a protein molecule are specified by __________.
   A. age
   B. sex
   C. heredity
   D. diet

5. What is meant by the amino acid sequence of a protein?
   A. Number of side chains in the protein
   B. Folding arrangement of the peptide chain
   C. Order of appearance of amino acids in the peptide chain
   D. Order of appearance of only the essential amino acids in the protein

6. Upon eating a hamburger, where does the breakdown of its proteins begin?
   A. Mouth
   B. Stomach
   C. Small intestine
   D. Large intestine

7. Which of the following is a protein catalyst which acts on other substances to change them chemically?
   A. Hormone
   B. Antibody
   C. Catalyst
   D. Enzyme

8. Which of the following is NOT a function of protein?
   A. Growth and maintenance of tissue
   B. Enzymes
   C. Antibodies
   D. Primary fuel for the brain

9. What is the relationship between body proteins and water?
   A. Proteins attract water
   B. Water attracts proteins
   C. Water degrades proteins
   D. Proteins form polymers of water

10. Negative nitrogen balance occurs in __________.
    A. surgery patients
    B. growing children
    C. pregnant women
    D. lactating women

11. What term describes the following quote: "The evil spirit that infects the first child when the second child is born"?
   A. Marasmus
   B. Kwashiorkor
   C. Psychomalnutrition
   D. Postbirth malnutrition

12. When amino acids are degraded for energy, their amine groups are stripped off and used elsewhere or incorporated by the liver into __________.
   A. bile
   B. urea
   C. glucose
   D. fat

13. Sickle cell disease is an example of an inherited mistake in the amino acid sequence.
   A. True
   B. False

14. Protein is needed for normal development of cells and muscle mass, but excessive amounts of protein __________.
   A. are even better and will bring faster results
   B. will not increase the building of body tissue
   C. will not add strength to the body
   D. B and C

15. Cardiovascular disease, colon cancer, and breast cancer risks have been shown to increase with diets high in __________.
   A. all protein
   B. plant protein
   C. animal protein
   D. carbohydrates

16. Marasmus can be characterized as a condition in which __________.
    A. overeating results in bloating and edema
    B. the person is starving from total malnutrition
    C. the immune system is impaired, allowing for microbe attack
    D. B and C

17. The primary component(s) of the School Lunch Program that help prevent protein-energy malnutrition in U.S. school children is (are) __________.
    A. meat or meat alternates
    B. milk
    C. pasta and breads
    D. A and B

18. For someone who has high blood cholesterol, protein intakes should __________.
    A. be less from animal sources and more from plant sources
    B. represent the bulk of the diet
    C. not be a concern if they are eating from the Food Guide Pyramid
    D. A and C

Essay Questions

Reflect on the items listed below; one or more may be included on the exam.

19. List the steps involved in protein synthesis.

20. Compare and contrast kwashiorkor and marasmus.

ANSWER KEY

The following provides the answers and references for the practice test questions.

	Answer	Learning Objectives	References
1.	D	LO 1	Sizer/Whitney, p. 180
2.	C	LO 2	Sizer/Whitney, p. 181
3.	A	LO 4	Sizer/Whitney, p. 181
4.	C	LO 4, 5	Sizer/Whitney, p. 183
5.	C	LO 4	Sizer/Whitney, p. 183
6.	B	LO 3	Sizer/Whitney, p. 186
7.	D	LO 6	Sizer/Whitney, p. 182
8.	D	LO 6	Sizer/Whitney, p. 188
9.	A	LO 6	Sizer/Whitney, p. 189
10.	A	LO 8	Sizer/Whitney, p. 197
11.	B	LO 9	Sizer/Whitney, p. 200
12.	B	LO 7	Sizer/Whitney, p. 191
13.	A	LO 5	Sizer/Whitney, p. 183
14.	D	LO 8	Video
15.	C	LO 10	Video
16.	D	LO 9	Video
17.	D	LO 9	Video
18.	D	LO 10	Video
19.		LO 4	Sizer/Whitney, p. 185
20.		LO 9	Sizer/Whitney, pp. 198–200

Lesson 8

The Protein Continuum

OVERVIEW

Because foods provide amino acids in differing amounts, it is difficult to know if the protein you eat has sufficient essential amino acids to support protein synthesis. How can you tell if the protein you eat is high quality? Which foods provide the best quality protein? How much is enough for optimum nutrition? In this lesson, you will look at how protein quality is determined. You will learn about the differences between food plans that include meat and food plans that do not contain meat. Understanding these differences will enable you to see how people can obtain enough protein from the diet regardless of their choice of protein sources.

LEARNING OBJECTIVES

Upon completing this lesson, you should be able to:

1. Distinguish among the following, using examples of foods: limiting amino acids, complete proteins, incomplete proteins, mutual supplementation, and complementary proteins.
2. Explain why protein digestibility-corrected amino acid scores (PDCAAS) are used in food labeling.
3. Calculate personal Dietary Reference Intakes (DRI) for protein and cite recommended intakes.
4. Describe the positive and adverse effects of protein and amino acid supplements.
5. Differentiate the health effects of vegetarian versus nonvegetarian eating plans.

LESSON ASSIGNMENT

Review the following assignment in order to schedule your time appropriately. Pay careful attention; the titles and numbers of the textbook chapter, the telecourse guide lesson, and the video program may be different from one another.

Text: Sizer and Whitney, *Nutrition: Concepts and Controversies*, Chapter 6, "The Proteins and Amino Acids," pp. 179–212, Appendix H

Video: "The Protein Continuum" from the series *Nutrition Pathways*

LESSON FOCUS POINTS

The following questions are designed to help you get the most out of the lesson. After you read the textbook assignments and view the video, you should be able to answer these questions.

1. How digestible are protein supplements compared to protein foods? What can happen when an excess of a single amino acid is eaten? What has research indicated for the amino acids lysine and tryptophan with regard to their use and effectiveness for treating specific conditions? (text, p. 193)

2. What is a limiting amino acid? What is a complete protein? What food sources generally provide complete proteins? What is an incomplete protein? What foods contain incomplete protein? To what does "complementary protein" refer? What is mutual supplementation? (text, pp. 194–196)

3. Provide three examples for each term listed in #2. (text, pp. 194–196)

4. What do PDCAAS measure? What is the maximum PDCAAS that a food can receive? Provide examples of foods with a rating of 1.00. Why are PDCAAS measures used as the basis for protein foods on food labels? (text, p. 196)

5. What is the recommended percent of total calories for protein? Why is the daily general DRI for protein intake based on grams per kilogram of body weight? What are the steps necessary to determine an individual protein DRI? (text, p. 197)

6. Define meat eaters. What effects do high-meat diets have on health? (text, p. 201; video)

7. What are reasons people choose to become vegetarians? What is a "semi-vegetarian"? (text, p. 207; video)

8. What are the positive health aspects of a vegetarian diet? What are the positive health aspects of a meat-based diet? What are possible adverse effects of a vegetarian diet? (text, pp. 207–212)

9. What nutrients might be missing from vegetarian diets? (text, pp. 210–211; video)

10. At what times during life do protein requirements increase? What is the typical protein intake for adults in the United States? (video)

11. What are the ways people can reduce daily protein intake? How can reducing protein intake affect weight loss? (video)

KEY CONCEPTS

Protein in Foods

Bread, 1 slice	2 g protein
Cooked cereal, rice, noodles, pasta, 1/2 c	3 g protein
Cheese, 1 1/2 oz	11 g protein
Egg, 1 large	6 g protein
Legumes, cooked, 1/2 c	7 g protein
Milk or yogurt, 1c	8 g protein
Poultry, beef, pork or lamb, 3 oz	26 g protein
Tofu, 3 oz	7 g protein
Cooked vegetables, 1/2 c	2 g protein

© 2000 Wadsworth/Thomson Learning

Mutual Supplementation

In general, legumes provide plenty of the amino acids isoleucine (Ile) and lysine (Lys), but fall short in methionine (Met) and tryptophan (Trp). Grains have the opposite strengths and weaknesses, making them a perfect match for legumes.

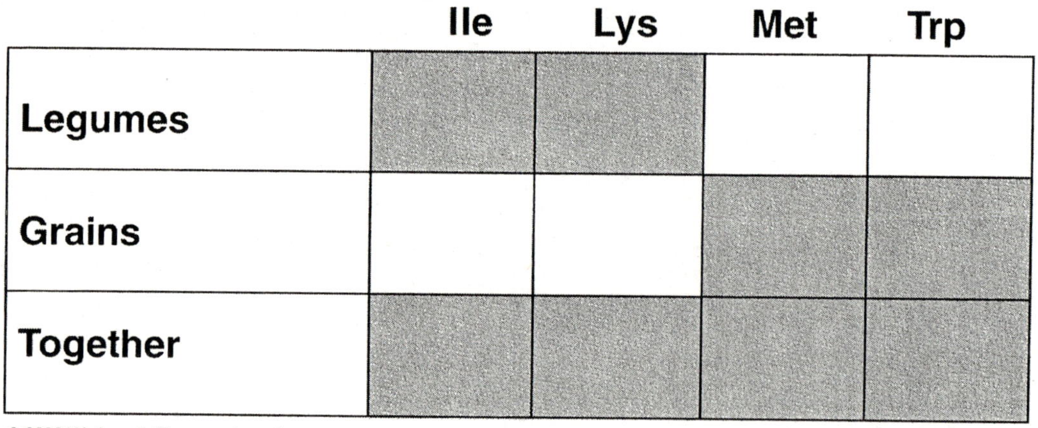

© 2000 Wadsworth/Thomson Learning

Protein Digestibility-Corrected Amino Acid Scores (PDCAAS) of Selected Foods

Food	PDCAAS[a]
Egg white	100
Ground beef	100
Chicken hot dogs	100
Milk protein (casein)	100
Fat-free milk powder	100
Beef salami	100
Tuna	100
Soybean protein	94
Whole-wheat and pea flour	82[b]
Pea flour	82[b]
Chick peas (garbanzos)	69
Kidney beans	68
Peas	67
Sausage, pork	63
Pinto beans	61
Rolled oats	57
Black beans	53
Lentils	52
Peanut meal	52
Whole wheat	40
Wheat protein (gluten)	25

a. Proteins with ideal digestion and amino acid balance are given a score of 100; others are scored against this standard.
b. An example of mutual supplementation. Combining whole wheat and pea flours yields a protein with a high PDCAAS than that of either product alone.

RELATED ACTIVITY

This activity is not required unless your instructor assigns it. It is offered as a suggestion to help you learn more about the material presented in this lesson.

Complete the form on this page and return it to your instructor according to the established deadline.

❑ Calculate Personal RDA for Protein

Calculate Personal RDA for Protein

The following exercise will help you to estimate your personal RDA for protein. Refer to Chapter 6 in your textbook for a sample calculation.

1. ___________ What is your weight in pounds?
2. ___________ Convert your weight to kilograms by dividing weight by 2.2. (Show your calculations below.)
3. ___________ Multiply kilograms by 0.8 to get your RDA in grams. Males younger than 18 years old, multiply by 0.9. (Show your calculations below.)
4. ___________ What is the RDA for protein for your age and sex (refer to the RDA table in Appendix H of your text)?
5. Answer the following questions pertaining to your personal protein RDA. Cite page numbers in your text to support your answers. Use the back if necessary.

   A. Explain whether you think the above calculated RDA (Step 3) for protein is adequate for you.

   B. List two examples of protein foods that you could include in your daily eating plan that would offer a high-quality, complete protein without cholesterol.
   1.
   2.

PRACTICE TEST

The following items will help you evaluate your understanding of this lesson. Use the answer key at the end of the lesson to check your answers or to locate material related in each question.

Multiple Choice

Select the one choice that best answers the question.

1. Which of the following is NOT considered to be a source of complete protein in human nutrition?
   A. Soy
   B. Egg
   C. Corn
   D. Fish

2. What is a limiting amino acid in a protein?
   A. A nonessential amino acid present in high amounts which inhibits protein synthesis
   B. An amino acid of the wrong structure to be utilized for protein synthesis efficiently
   C. An essential amino acid present in insufficient quantity for body protein synthesis to take place
   D. An amino acid that limits the absorption of other essential amino acids by competing with them for transport sites within the GI tract

3. Which of the following foods has the best assortment of essential amino acids for the human body?
   A. Eggs
   B. Fish
   C. Corn
   D. Rice

4. Which of the following groups of the Food Guide Pyramid contribute an abundance of high-quality protein?
   A. Milk
   B. Meat
   C. Vegetable
   D. A and B

5. For healthy adults, the Dietary Reference Intake for protein is set at ________.
   A. 0.8 gram per kilogram of body weight
   B. 2.2 pounds per kilogram of body weight
   C. 12 to 15 percent of total calories
   D. 100 grams per day

6. What is the DRI for protein for a 48 kilogram woman?
   A. 24 grams
   B. 34 grams
   C. 38 grams
   D. 40 grams

7. All of the following are advantages of vegetarian diets EXCEPT __________.
   A. fat content is lower
   B. fiber content is higher
   C. vitamin B_{12} intake is higher
   D. vitamins A and C are found in liberal quantities

8. Examples of complementary protein combinations include all of the following EXCEPT __________.
   A. pasta with tomato sauce
   B. rice and black-eyed peas
   C. broccoli with walnuts
   D. peanut butter and jelly on whole-wheat bread

9. Of the following foods, which has the highest PDCAAS?
   A. Chick peas
   B. Soybean protein
   C. Kidney beans
   D. Tuna

10. Protein intakes may need to increase as a result of illness as well as ________.
   A. when women are lactating
   B. during moderate physical activity
   C. from age two to five years
   D. any of the above

11. People who consume a diet in which meat is the primary source of protein are ________.
   A. at less risk for cancer than vegetarians
   B. thought to be at greater risk for cancer than vegetarians
   C. more likely to develop heart disease than vegetarians
   D. B and C

12. Recommendations from experts regarding a vegetarian diet are that ________.
   A. vegan pregnant/lactating women should take a B_{12} supplement
   B. lacto-ovo vegetarians might need an iron supplement
   C. vegetarians eat fewer calories and tend to live longer
   D. all of the above

13. For people who wish to lose weight, experts might advise them to ________.
   A. consider a more vegetarian eating plan for protein choices
   B. stay at the top of the Food Guide Pyramid for food choices
   C. focus on the middle of the Food Guide Pyramid for choices
   D. eat only plant-based proteins, completely eliminating meat

Essay Question

Reflect on the items listed below; one or more may be included on the exam.

14. Why would you advise someone NOT to take amino acid supplements?

15. List the health advantages of a vegetarian diet. What nutrient requirements are more difficult to meet on this diet, and what precautions are needed to prevent insufficient intakes in the child and in the adult?

ANSWER KEY

The following provides the answers and references for the practice test questions.

Answer		Learning Objectives	References
1.	C	LO 1	Sizer/Whitney, p. 196
2.	C	LO 1	Sizer/Whitney, p. 194
3.	A	LO 2	Sizer/Whitney, p. 196
4.	D	LO 1	Sizer/Whitney, p. 197
5.	A	LO 3	Sizer/Whitney, p. 196
6.	C	LO 3	Sizer/Whitney, pp. 196–197
7.	C	LO 5	Sizer/Whitney, pp. 207–210
8.	A	LO 1	Sizer/Whitney, p. 195
9.	D	LO 2	Sizer/Whitney, p. 196
10.	A	LO 3	Video
11.	D	LO 5	Video
12.	D	LO 5	Video
13.	A	LO 5	Video
14.		LO 4	Sizer/Whitney, p. 193
15.		LO 5	Sizer/Whitney, pp. 207–212

Lesson 9

Metabolism

OVERVIEW

Now that you know how the food you eat is digested, absorbed, and transported, how does the body actually obtain the energy it needs from that digested food? This is where metabolism comes into play. Metabolism is often the topic of conversation in health clubs and gyms, for body builders, athletes, and others who are interested in controlling weight. Yet, to most people not involved in the biological sciences, it is one of the most confusing—and overwhelming—physiological phenomena associated with the human condition.

In this segment, you will examine how food consumed in excess is converted to body fat. Along with feasting, the effects of fasting and starvation on the body will also be examined, as well as the effects of alcohol on metabolic functioning.

LEARNING OBJECTIVES

Upon completing this lesson, you should be able to:

1. Explain the effects that feasting has on metabolism.

2. Explain the effects of short-term fasting on metabolism.

3. Describe the metabolic process during long-term fasting/starvation.

4. Distinguish between the positive and adverse health effects of alcohol on nutrition status.

LESSON ASSIGNMENT

Review the following assignment in order to schedule your time appropriately. Pay careful attention; the titles and numbers of the textbook chapter, the telecourse guide lesson, and the video program may be different from one another.

Text: Sizer and Whitney, *Nutrition: Concepts and Controversies*, Chapter 9, "Energy Balance and Healthy Body Weight," pp. 311–357

Video: "Metabolism" from the series *Nutrition Pathways*

LESSON FOCUS POINTS

The following questions are designed to help you get the most out of the lesson. After you read the textbook assignments and view the video, you should be able to answer these questions.

1. After about ten days of fasting, what is the brain's alternate fuel source? What is ketosis? What happens to the metabolism during a fast? What happens to lean body losses and fat losses during a fast? List the physical symptoms of starvation. (text, pp. 330–332)

2. Name the basic units derived from the carbohydrates, protein, and fat in food. What is the body's preferred energy-storage form? What happens to excess dietary fat if the body doesn't need more energy? What happens to excess dietary protein if it is not needed for building tissues or for energy? What is the first compound stored from excess carbohydrate intake? What happens when glycogen stores are filled? (text, pp. 331–332)

3. What is the first thing to happen to amino acids if they are needed for energy or eaten in excess? At the beginning of a fast, what fuel is most depleted? What is the preferred fuel for the brain, red blood cells, and the nervous system? (text, pp. 331–332)

4. How does alcohol promote fat storage? How does long-term excessive alcohol intake contribute to malnutrition? How does moderate alcohol intake impact health? What is moderate alcohol intake for a woman? For a man? (text, p. 333)

5. How is metabolism like a power plant? (video)

6. What nutrient on a weight-to-weight basis is the most calorie dense? (video)

7. How can physical activity affect calorie expenditure and metabolism? How does extra body fat affect the circulatory system? (video)

8. What is the effect of fasting on the brain and heart? (video)

9. How does alcohol affect metabolism? (video)

KEY CONCEPTS

Feasting

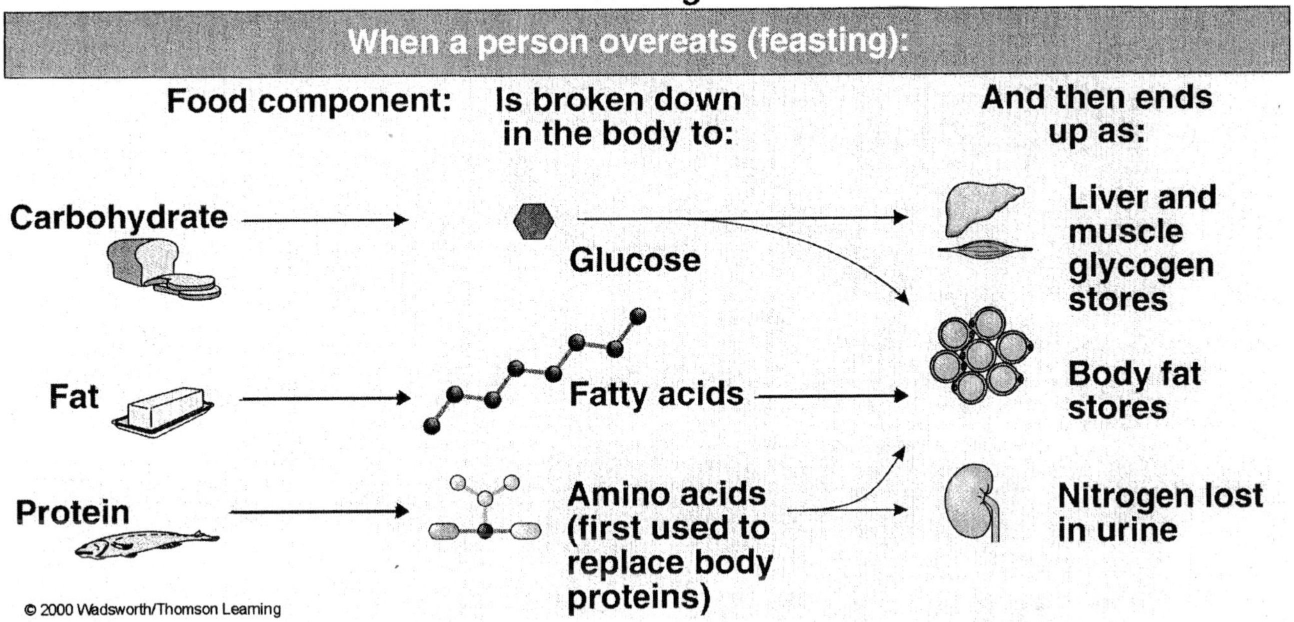

Fasting

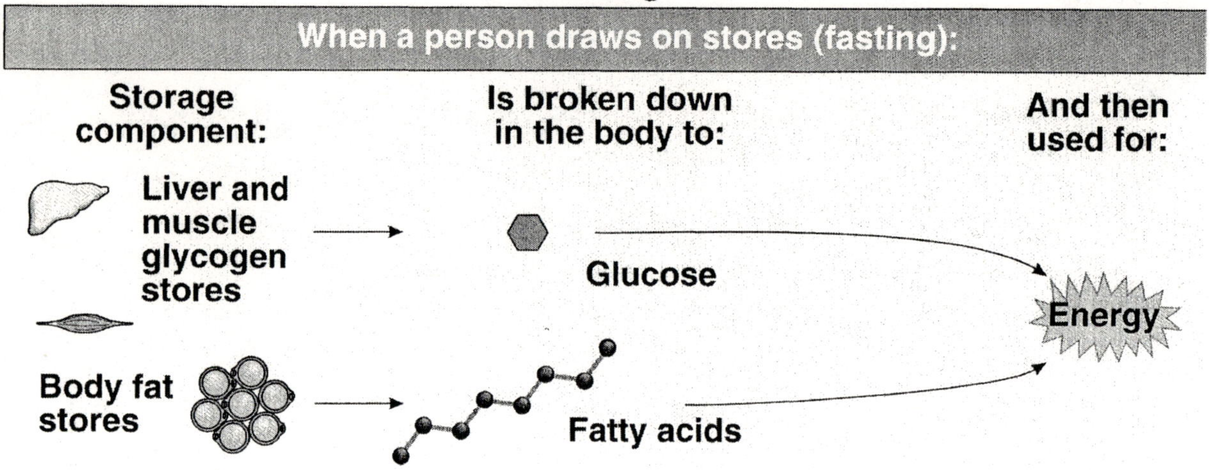

© 2000 Wadsworth/Thomson Learning

Continued Fasting

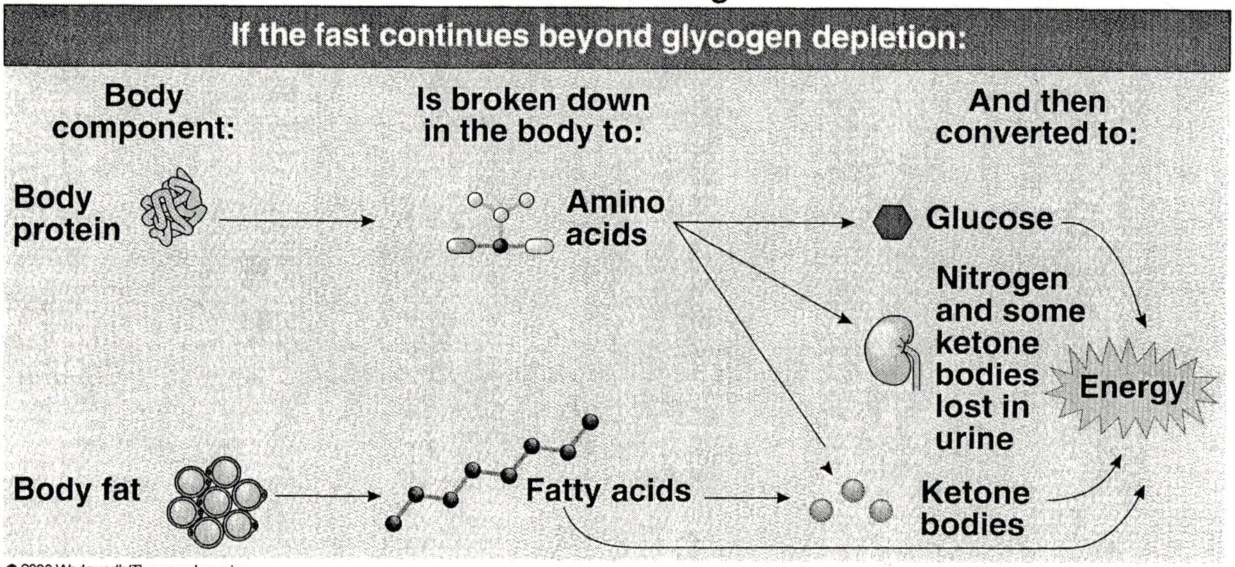

© 2000 Wadsworth/Thomson Learning

PRACTICE TEST

The following items will help you evaluate your understanding of this lesson. Use the answer key at the end of the lesson to check your answers or to locate material related in each question.

Multiple Choice

Select the one choice that best answers the question.

1. Quick, large changes in weight are most likely the result of all of the following EXCEPT changes in __________.
   A. fat
   B. body fluid content
   C. bones' minerals
   D. lean tissues such as muscles

2. Any low-calorie diet is accompanied by __________.
   A. lean tissue loss
   B. ketosis
   C. loss of appetite
   D. fast weight loss

3. A healthy person starting with average body fat can live totally deprived of food for as long as six to eight weeks due to __________.
   A. acidosis
   B. thermogenesis
   C. metabolism
   D. ketosis

4. In early food deprivation, the nervous system cannot use _____ as fuel.
   A. fat
   B. glucose
   C. protein
   D. carbohydrate

5. Physiological hazards of low-carbohydrate diets include __________.
   A. heart failure
   B. hyperglycemia
   C. high blood cholesterol
   D. A and C

6. Which of the following is NOT a true statement?
   A. Carbohydrate from food is especially easy for the body to store as fat.
   B. Protein is stored in the body in response to exercise.
   C. Alcohol delivers calories and encourages fat storage.
   D. Any food can make you fat if you eat enough of it.

7. When protein consumption is in excess of body needs and energy needs are met, the excess amino acids are metabolized and the energy in the molecules is __________.
   A. stored as fat only
   B. excreted in the feces
   C. stored as amino acids only
   D. stored as glycogen and fat

8. The three kinds of energy nutrients can be stored in the body as __________.
   A. fat
   B. water
   C. glycogen
   D. A and C

9. After the first day or so of fasting, which of the following is most depleted in the body?
   A. Glycogen
   B. Fatty acids
   C. Amino acids
   D. Triglycerides

10. How soon would death occur from starvation if the body was unable to shift to a state of ketosis?
    A. Within three weeks
    B. Less than two weeks
    C. Between five and six weeks
    D. Between two and three months

11. Which of the following dietary nutrients would most rapidly reverse a state of ketosis in a starving person?
    A. Fat
    B. Protein
    C. Amino acids
    D. Carbohydrate

12. A power plant and the body's metabolism are similar in that both __________.
    A. generate energy
    B. store unused energy
    C. release energy when needed
    D. all of the above

13. Every time you put on an extra pound of body fat, you can __________.
    A. increase blood pressure
    B. decrease blood pressure
    C. become more susceptible to some cancers
    D. A and C

Fill in the Blank

Insert the correct word or words in the blank for each item.

14. The nutrient that is the most calorie dense on a weight-to-weight basis is ______________.

15. Long-term fasting can cause the heart muscle to ________________.

16. Alcohol intake impacts the body's use of B vitamins by ________________ ____________________.

Essay Questions

Reflect on the items listed below; one or more may be included on the exam.

17. Discuss ways in which the body's metabolism adapts to conditions of fasting/starvation. How do these adaptations affect the rate of weight loss when dieting?

18. What is ketosis and how can it be identified? What conditions typically induce a state of ketosis? What are the adverse effects of this abnormality?

ANSWER KEY

The following provides the answers and references for the practice test questions.

	Answer	Learning Objectives	References
1.	A	LO 2	Sizer/Whitney, p. 329
2.	C	LO 3	Sizer/Whitney, p. 332
3.	D	LO 3	Sizer/Whitney, p. 330
4.	A	LO 2	Sizer/Whitney, p. 330
5.	D	LO 3	Sizer/Whitney, p. 332
6.	A	LO 4	Sizer/Whitney, pp. 337–338
7.	D	LO 1	Sizer/Whitney, p. 332
8.	D	LO 1	Sizer/Whitney, p. 332
9.	A	LO 2	Sizer/Whitney, p. 330
10.	B	LO 3	Sizer/Whitney, p. 330
11.	D	LO 3	Sizer/Whitney, p. 330
12.	D	LO 1, 2, 3	Video
13.	D	LO 1	Video
14.	fat (fatty acids)	LO 1	Video
15.	diminish in size (get smaller)	LO 3	Video
16.	preventing absorption (depleting vitamins)	LO 4	Video
17.		LO 2, 3	Sizer/Whitney, pp. 329–332
18.		LO 3	Sizer/Whitney, pp. 329–332

Lesson 10

Weight Control: Energy Regulation

OVERVIEW

The food you eat is not only converted to energy for the present, but it is also converted to energy for future needs—mainly in the form of body fat which virtually has an unlimited capacity for storage. It seems amazing, considering the daily intake of food and the daily expenditure of energy from that food, that most of us are able to maintain a fairly stable body weight from year to year.

In this lesson, you will explore the complexities of weight control by discussing energy balance, body weight, body composition, and the ways health is affected by overweight and underweight. You will begin by looking at how the body expends the energy it obtains from food through basal metabolism, physical activity, thermic effect, and adaptive thermogenesis. You will estimate your personal energy requirements and begin to understand how your own metabolism affects weight gain, loss, or maintenance. The importance of healthy body weight versus the distorted view of the "ideal" body presented by the media will be discussed. Finally, you will learn the best ways to determine healthy body composition and the factors for good health that must be taken into account beyond a bathroom scale.

LEARNING OBJECTIVES

Upon completing this lesson, you should be able to:

1. Describe energy balance.
2. Explain the three components of energy expenditure and factors that affect them.
3. Cite the health risks associated with being underweight, being overweight, and central obesity.
4. Discuss the role of standard weight tables and body mass index (BMI) in defining obesity.

5. Describe the methods of determining body composition.

6. List body fat percentages for normal-weight adults.

7. Identify the factors that affect food intake.

8. Describe the influence of the media on body image.

LESSON ASSIGNMENT

Review the following assignment in order to schedule your time appropriately. Pay careful attention; the titles and numbers of the textbook chapter, the telecourse guide lesson, and the video program may be different from one another.

Text: Sizer and Whitney, *Nutrition: Concepts and Controversies*, Chapter 9, "Energy Balance and Healthy Body Weight," pp. 311–357

Video: "Weight Control: Energy Regulation" from the series *Nutrition Pathways*

LESSON FOCUS POINTS

The following questions are designed to help you get the most out of the lesson. After you read the textbook assignments and view the video, you should be able to answer these questions.

1. What health risks are associated with being underweight? Explain the relationship between overweight and the following: cardiovascular disease, type 2 diabetes, and cancer. How does fat distribution relate to increased risk for disease? What is "central" obesity? What is meant by "apples" and "pears" relative to fat distribution? (text, pp. 312–313)

2. How many calories are in one pound of body fat? How many calories per day must a person expend or take in to lose or gain one pound of weight in a month? (text, p. 316)

3. What is basal metabolism? How much energy spent in a day comes from basic metabolic functions? What factors raise basal metabolic rate (BMR)? What factors lower BMR? What are the two major ways in which the body spends

energy? Define the thermic effect of food. How much energy per day comes from the thermic effect of food? (text, p. 317)

4. What components of energy expenditure are considered when calculating energy RDA? To estimate energy spent on BMR, what factors are used? What steps are taken to estimate energy output for an individual? (text, pp. 317–319)

5. What are the criticisms against height and weight tables as criteria for health risks? Define *body mass index*. What BMI is associated with overweight and obesity? What BMI is associated with the lowest risk to health for younger and older adults? What are the drawbacks of using weight measures in assessing disease risk? (text, pp. 318–322)

6. What is the most important factor for determining how much a person should weigh or how much body fat a person should carry? What does a fatfold measure estimate? What other methods are used to assess body composition? (text, pp. 318–322)

7. What is the percent body fat range for a normal-weight man and for a normal-weight woman? (text, p. 322)

8. What is the difference between hunger and appetite? What is satiety? Cite examples of eating habits that override the signals of hunger and satiety. What organ is the control center for eating behavior? (text, pp. 323–325)

9. Identify factors that affect food intake other than appetite and hunger. What should internal cues regarding food intake tell people? (text, pp. 327–328; video)

10. Who is the primary target of media with regard to ideal body image and size? How has media contributed to the dieting craze in the United States? (video)

11. Define "desirable" body weight. What is good advice regarding a person's body size? (video)

12. What factor has the greatest influence on maintaining weight loss? (text, p. 342; video)

13. How does a lower metabolic rate impact the ability to lose weight successfully? (video)

KEY CONCEPTS

Medical Problems Associated with Obesity

- Hypertension
- Renal disease
- Gallbladder disease
- Pulmonary disease
- Problems with anesthesia during surgery
- Osteoarthritis and gout
- Breast and endometrial cancer
- Abnormal plasma lipid and lipoprotein concentrations
- Impairment of cardiac function
- Menstrual irregularities and toxemia of pregnancy
- Psychological trauma
- Flat feet and infection of fatfolds
- Organ compression by adipose tissue
- Impaired heat tolerance

Source: Art Gilbert, Director, Wellness Fitness Institute, University of California, Santa Barbara, with permission.

Enrichment BMI silhouettes

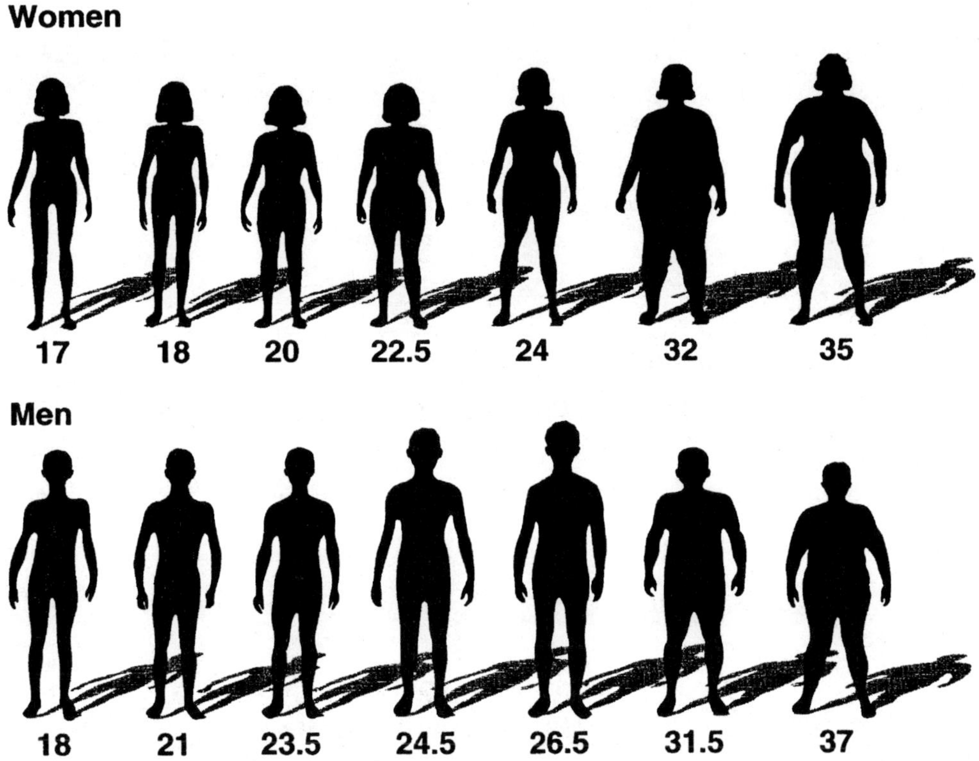

© 2000 Wadsworth/Thomson Learning

BMI Values and Health Risks

BMI	Obesity Class	Risks to Health
<18.5	Underweight	The lower the BMI, the greater the risk
18.5 through 24.9	Normal	Very low risk
25.0 through 29.9	Overweight	Increased risk/ high risk
30.0 through 34.9	Class I obesity	High risk/very high risk
35.0 through 39.9	Class II obesity	Very high risk
40.0 or above	Class III obesity	Extreme risk

© 2000 Wadsworth/Thomson Learning

RELATED ACTIVITY

This activity is not required unless your instructor assigns it. It is offered as a suggestion to help you learn more about the material presented in this lesson.

Complete the form on this page and return it to your instructor according to the established deadline.

❑ How Many Calories Do I Need Each Day?

How Many Calories Do I Need Each Day?

This exercise will help you estimate the number of calories per day you need to maintain your weight.

1. ___________ Your current (or preferred) weight in pounds

2. ___________ Convert pounds to kilograms by dividing your body weight by 2.2. (Show your calculations below.)

3. ___________ Multiply weight in kilograms by the BMR factor: 1 cal/kg/hr. 1 cal/kg/hr x _____ kg = _____ x 24 hrs/day = _____ calories per day. This number represents your total daily energy need required to maintain the weight indicated in Step 1.

4. ___________ Do you feel the above estimation accurately reflects the number of calories per day you need to maintain the weight indicated in Step 1? Why or why not? Cite page numbers in textbook to support your answer.

PRACTICE TEST

The following items will help you evaluate your understanding of this lesson. Use the answer key at the end of the lesson to check your answers or to locate material related in each question.

Multiple Choice

Select the one choice that best answers the question.

1. Which of the following is NOT a health risk associated with deficient body fat?
   A. Increased risk during hospitalization
   B. Increased risk of developing hypertension
   C. Increased risk for individuals fighting a wasting disease
   D. Increased chance of death from cancer

2. What would be the approximate weight gain of a person who consumes an excess of 500 calories daily for one month?
   A. Half a pound
   B. Two pounds
   C. Three pounds
   D. Four pounds

3. One pound of body fat is equal to ______ calories.
   A. 2,500
   B. 3,000
   C. 3,500
   D. 4,000

4. The psychological desire to eat is referred to as __________.
   A. satiety
   B. appetite
   C. hunger
   D. obesity

5. Which of the following is a feature of the basal metabolic rate (BMR)?
   A. Fever decreases the BMR
   B. Fasting increases the BMR
   C. Pregnancy increases the BMR
   D. Females have higher BMRs than males on a body weight basis

6. If a dancer and a typist are the same height and have the exact same body build, the dancer will be heavier because she has __________.
   A. more body fat
   B. stronger bones
   C. stronger muscles
   D. more muscle mass

7. You are weighing a team of football players, and according to the BMI values they appear to be obese. Which of the following would you most likely conclude?
   A. The majority of them are probably obese and should make efforts to lose weight.
   B. Their muscle weight is probably responsible for the elevated scale weights.
   C. Their bones may be well mineralized, which contributes to their weights.
   D. B and C

8. Paul is a nineteen-year-old athlete who is 6' tall and weights 178 pounds; Rick is a forty-year-old salesman who weighs 165 pounds and is 5'7" tall. Who would have the higher basal metabolic rate?
   A. Rick
   B. Paul

9. Which of the following defines central obesity?
   A. Accumulation of fat during the mid-years of life
   B. Storage of excess fat around the central part of the body
   C. Overfatness due to a large number of interacting behavioral problems
   D. Overfatness due to reliance on high fat foods as a central part of the diet

10. Excess body fatness is associated with an increased risk of __________.
   A. some cancers
   B. gallbladder disease
   C. diverticular disease
   D. A and B

11. Factors that affect food intake include not only hunger and appetite, but also include __________.
    A. genetics
    B. the drive for energy (calories)
    C. age
    D. all of the above

12. What population has been the primary target of body image idealization especially in the U.S. culture?
    A. Males
    B. Adolescents
    C. Females
    D. Children

13. How has the media contributed to the dieting phenomena in the United States?
    A. By idealizing the average housewife
    B. By idealizing the average athlete
    C. By idealizing female models
    D. By idealizing male models

14. Desirable weight might be defined as that weight which __________.
    A. is not associated with hypertension, diabetes, or heart disease
    B. is associated with the lowest mortality (death) rate
    C. can be maintained for more than five years
    D. A and B

15. In studies performed on women who were able to maintain their lost weight, findings showed that __________.
    A. they attended regular support groups
    B. they kept regular track of food intake
    C. 90 percent of them exercised regularly
    D. all of the above

16. If people are trying to lose weight and are 20 percent below the predicted equation for size, sex, and age for metabolic rate, they will __________.
    A. have a harder time burning calories and not lose weight
    B. be tired most of the time and gain weight
    C. expect to lose weight quickly on 1,500 calories per day
    D. A and B

Essay Questions

Reflect on the items listed below; one or more may be included on the exam.

17. How would you respond to an advertisement for a weight-loss diet that claims that eating certain foods can elevate the BMR and thus promote weight loss? How can someone change her BMR?

18. Compare and contrast methods of measuring body composition. Identify three groups of people who are more prone to develop central obesity. Why is fat that collects in the central abdominal area of the body especially dangerous?

ANSWER KEY

The following provides the answers and references for the practice test questions.

	Answer	Learning Objectives	References
1.	B	LO 3	Sizer/Whitney, p. 312
2.	D	LO 1	Sizer/Whitney, p. 316
3.	C	LO 1	Sizer/Whitney, p. 316
4.	B	LO 7	Sizer/Whitney, p. 323
5.	C	LO 2	Sizer/Whitney, p. 317
6.	D	LO 6	Sizer/Whitney, p. 318
7.	D	LO 4	Sizer/Whitney, p. 320
8.	B	LO 2	Sizer/Whitney, pp. 317–318
9.	B	LO 3	Sizer/Whitney, p. 313
10.	D	LO 3	Sizer/Whitney, p. 313
11.	D	LO 7	Video
12.	C	LO 8	Video
13.	C	LO 8	Video
14.	D	LO 3	Video
15.	C	LO 2	Video
16.	A	LO 2	Video
17.		LO 2	Sizer/Whitney, pp. 316–317
18.		LO 5	Sizer/Whitney, pp. 313 and 319-320

Lesson 11

Weight Control: Health Effects

OVERVIEW

You just can't stand it anymore. Your clothes don't fit like they used to, and you just don't have energy anymore. You know you need to lose the weight and with all the diet programs, pills, potions, and gadgets out there it will be a snap to lose those extra pounds—or will it? How do you know the weight will stay off once you've lost it? What are the best weight-loss methods to use? Should you sign up for one of the commercial diet programs you see advertised on TV or go to a weight-loss clinic? What should you look for in a sound weight-loss program? These questions and more will be the focus of this lesson as you examine the treatment and health effects of weight control.

In addition to learning about strategies for achieving a healthy body weight and extreme treatments for obesity, you will also look at the causes and treatments for the eating disorders of anorexia nervosa and bulimia nervosa, and gain insight as to why certain people become victims of these disorders.

LEARNING OBJECTIVES

Upon completing this lesson, you should be able to:

1. Summarize the theories that attempt to explain the mystery of obesity.
2. Describe controversies in obesity treatment.
3. Explain extreme treatment choices for weight loss including surgery and pills.
4. Discuss current recommended strategies for weight loss.
5. Describe strategies for gaining weight.
6. Differentiate between anorexia nervosa and bulimia nervosa.
7. Compare fashion models to average U.S. women in body size and/or body weight.
8. Explain how waist-to-hip ratios affect risk to health.

LESSON ASSIGNMENT

Review the following assignment in order to schedule your time appropriately. Pay careful attention; the titles and numbers of the textbook chapter, the telecourse guide lesson, and the video program may be different from one another.

Text: Sizer and Whitney, *Nutrition: Concepts and Controversies*, Chapter 9, "Energy Balance and Healthy Body Weight," pp. 311–357

Video: "Weight Control: Health Effects" from the series *Nutrition Pathways*

LESSON FOCUS POINTS

The following questions are designed to help you get the most out of the lesson. After you read the textbook assignments and view the video, you should be able to answer these questions.

1. Explain how each of the following can contribute to being overweight: fat cell development, fat cell metabolism, and set-point theory. Explain how genetics, overeating, and physical inactivity may cause obesity. What is leptin? What is lipoprotein lipase? Which contributors can people control? (text, pp. 324–327)

2. Describe the assumptions and science concerning high protein diets. Define "yo-yo" dieting. What is weight cycling? What are the risks associated with yo-yo dieting? Explain how ineffective weight-loss treatment can add to a dieter's psychological burden. Describe how the following drugs function as weight-loss aids: diuretics, amphetamines, fenfluramine, phentermine, phenylpropanolamine, and herbal products. Describe "gastric bypass" and "gastric banding" as weight-loss methods. (text, pp. 334–335; 344–347)

3. State the prejudices against overweight people. How have obesity experts revised their thinking with regard to reasonable treatments for obesity? What parameters are more useful than body weight in denoting successful weight loss? What do experts consider to be a "reasonable goal weight"? (text, pp. 336–339)

4. What is considered a reasonable time frame for a 10 percent loss of initial weight? What are the only considerations that need to be taken into account when designing a personal eating plan? What is the main characteristic of a

weight-loss diet? What is the lower limit of calories per day an adult should consume to provide adequate nutrition when trying to lose weight? What is the goal with regard to food portions? What should be the center of meals and snacks? What is the relationship of fat loss to exercise and dieting? Describe the effect of activity on the following: BMR, appetite, psychological well-being, and spot reducing. What type of activity is most effective for weight control? (text, pp. 336–343)

5. State strategies for weight gain. (text, pp. 343–344)

6. Define gastric bypass. What population are candidates for gastric bypass procedures? (text, pp. 345–346; video)

7. Describe the health effects of "yo-yo" dieting. (text, p. 347; video)

8. State behavior techniques or strategies that people can use to help with a weight-loss program. (text, pp. 348–349)

9. Explain the three components of the Female Athlete Triad. Describe the differences between anorexia and bulimia with regard to the following: cause, diagnosis, symptoms, and treatment. Differentiate between bulimia and binge-eating disorder. What are the criteria for diagnosis of binge-eating disorder? (text, pp. 351–357)

10. Which eating disorder is most medically dangerous? What are criteria for diagnosing and treating anorexia and bulimia? What are causes of eating disorders? (text, pp. 351–357; video)

11. How do average women in the United States compare in body weight and shape to fashion models? (video)

12. What are the criteria for successful weight loss? (video)

13. What are the health effects of high waist-to-hip ratios? How does having an "apple" shape impact risk for disease? (video)

14. What is considered the most sound advice for successful weight loss? What advice should be given to women who have gained weight during pregnancy? (video)

A Model of Leptin's Effects on Energy Balance and Body Fatness

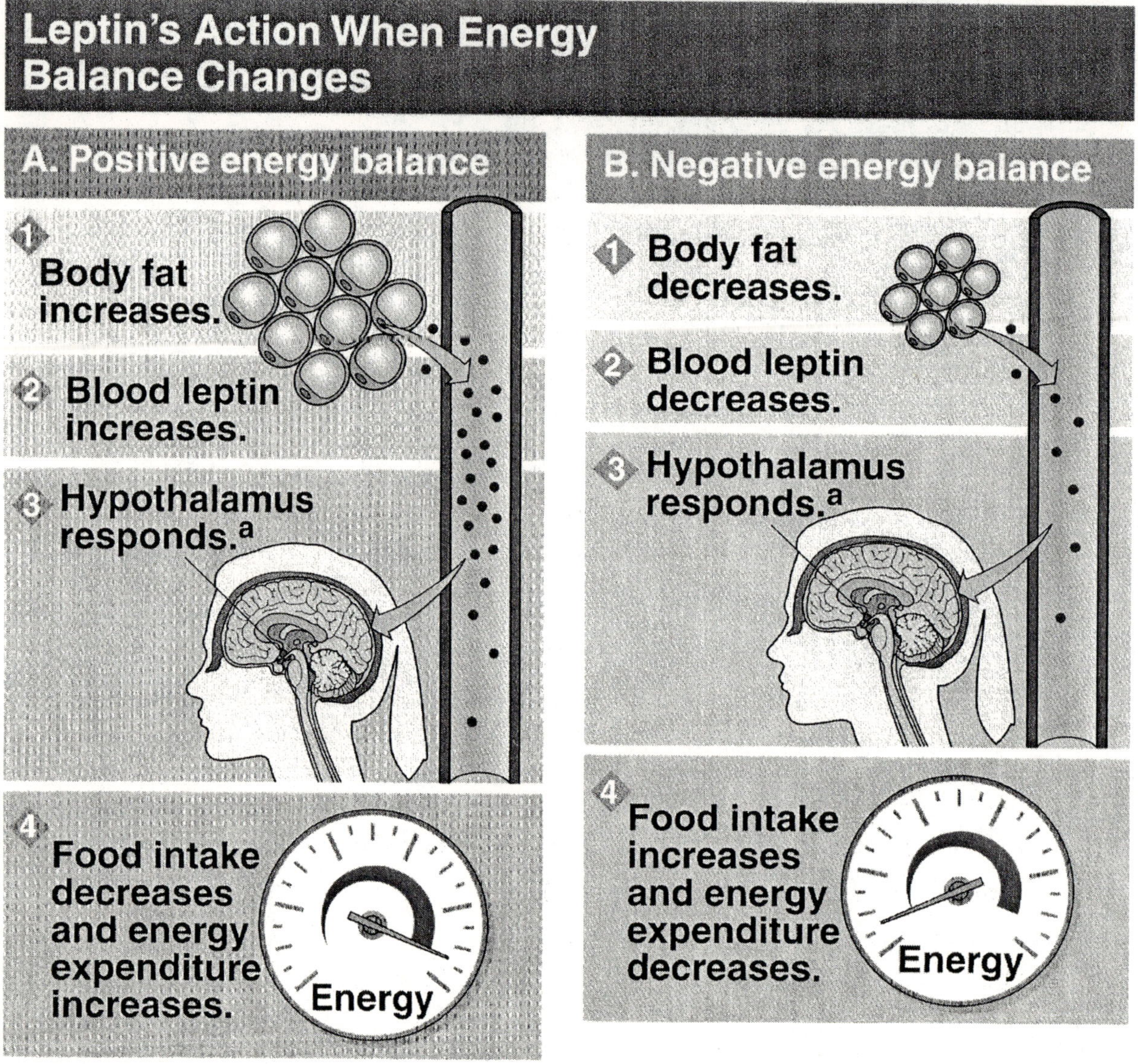

[a] The brain's hypothalamus was introduced in Chapter 3.

© 2000 Wadsworth/Thomson Learning

Weight Cycling Effect of Repeated Dieting

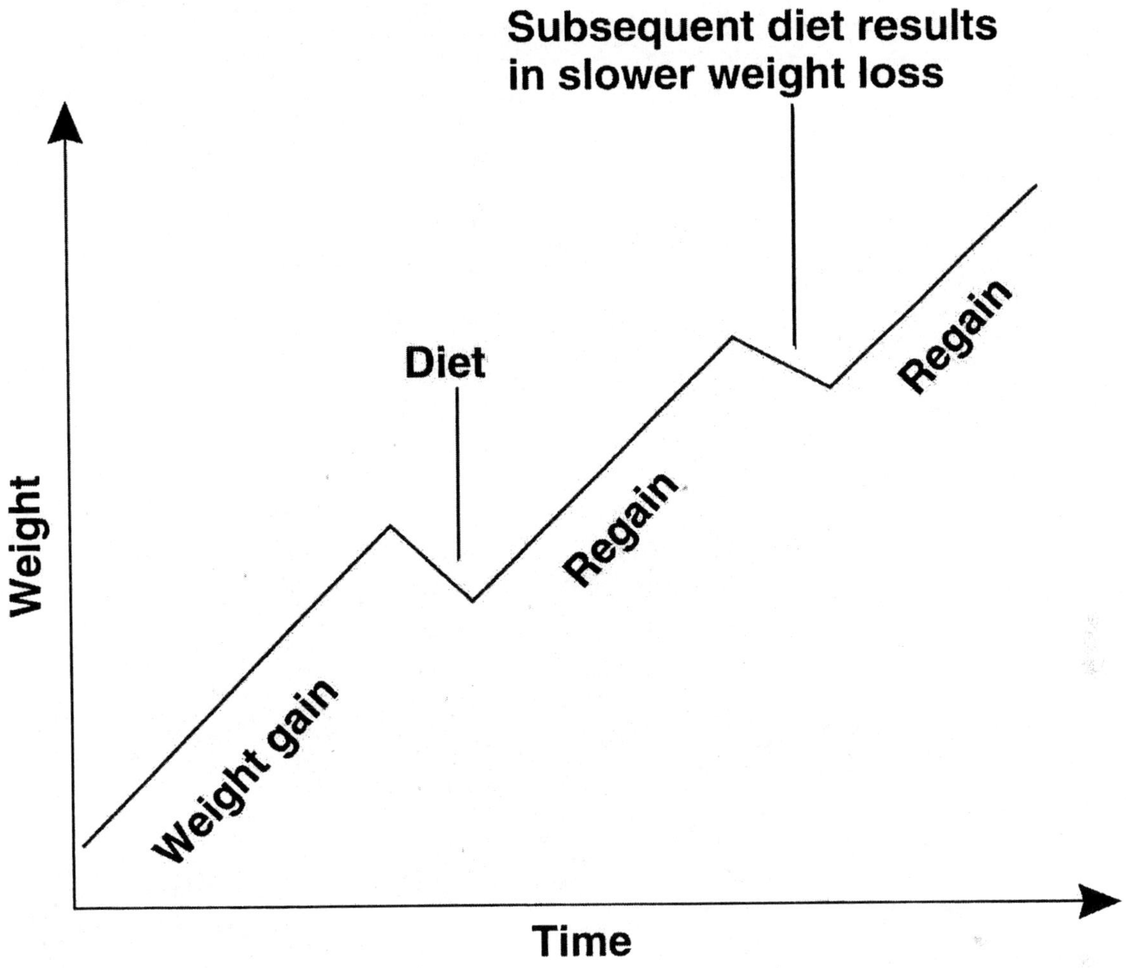

© 2000 Wadsworth/Thomson Learning

Suggested Daily Servings for an Adequate 1,200-Calorie Diet

Choose the lowest-calorie, fat-free, or lowest-fat options from each group. Strictly limit serving sizes to those specified in Figure 2-7 of Chapter 2.

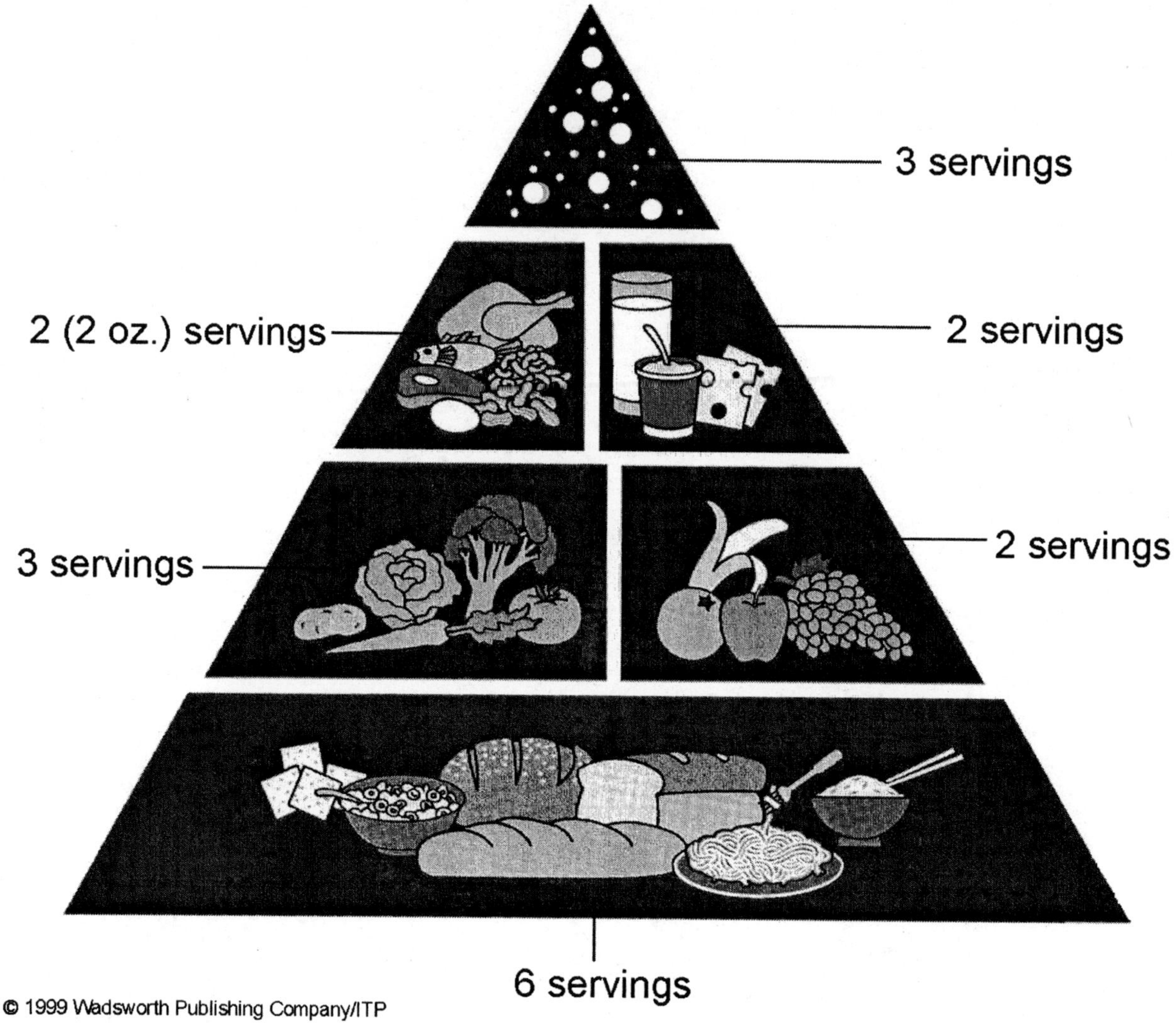

© 1999 Wadsworth Publishing Company/ITP

*Assumes no alcohol intake

Activities and Rewards to Substitute for Eating

- Attending sporting events
- Enjoying leisure activities
- Exercising or playing sports
- Gardening
- Getting praise from others
- Going to a movie or play
- Listening to music
- Napping
- Praising yourself
- Reading
- Receiving token rewards (stickers, seals)
- Redecorating
- Relaxing
- Saving money for future treats
- Shopping
- Taking a bubble bath
- Telephoning
- Tidying your room or house
- Vacationing
- Working on hobbies or crafts

RELATED ACTIVITIES

These activities are not required unless your instructor assigns them. They are offered as suggestions to help you learn more about the material presented in this lesson.

Complete the attached forms for the following related activities and return them to your instructor according to the established deadline.

- ❑ Designing a Personal Weight Control Plan
- ❑ Evaluation of a Commercial Weight-Loss Program

Designing a Personal Weight Control Plan

The purpose of this activity is to help you create an eating and exercise plan that will allow you to lose, gain, or maintain your weight based on health and your personal goals. Use diet analysis software and information in Chapter 9 to complete this activity, following the steps below.

Step 1: Examine your original three-day food intake from your Personal Dietary Analysis Project. Each of the three days should meet the following standards:

- At least 1,200 calories per day to maintain metabolism and prevent nutrient deficiencies.
- Greater than 50 percent of total calories from carbohydrate to maintain glucose and spare protein.
- Less than 30 percent of total calories from fat to help reduce the risk for chronic diseases.
- Between 10 and 20 percent of total calories from protein for growth, fluid balance, enzymes, etc.
- The *minimum* recommended servings for each level of the Food Guide Pyramid to ensure adequacy, balance, and nutrient density.

Step 2: If your original three days did NOT meet the above criteria, then you need to make the necessary modifications in one or more of the following ways:

- Adjust serving sizes up or down.
- Substitute low- or non-fat foods for higher fat foods (to reduce total fat, saturated fat, and/or cholesterol intake).
- Substitute plant proteins for animal proteins to reduce fat and/or cholesterol intake (see Related Activity in Lesson 8 for suggestions).
- Add foods to reach the minimum number of servings for each level of the Food Guide Pyramid.

Step 3: Print the following information from the diet analysis software program:

- The original three-day *average* and Food Guide Pyramid
- The modified three-day *average* and Food Guide Pyramid
  (Please DO NOT print individual days—save a tree!)

Step 4: Answer the questions that follow, then:

- Staple the questions and answers to the above printouts
- Submit to your professor by the stated deadline

1. Record your current weight and your desired weight, if different.

   ______________ Current weight

   ______________ Desired weight

2. Select one of the following goals based on the weight that is *healthiest* for you:

   ______________ Weight maintenance

   ______________ Weight loss

   ______________ Weight gain

3. ______________ How many calories per day will you consume to reach the above weight goal? (Note: When modifying your eating plan, stay within +/- 250 calories per day of the amount you indicate here.)

4. Briefly explain what general modifications were made to your *original* three-day food intake.

5. List the foods you added (or deleted) along with their serving sizes for each level of the Pyramid.

   Grains, cereals, breads, etc:

   Fruits:

   Vegetables:

   Meat/Meat alternates:

   Dairy:

6. Describe an exercise plan designed by you that would help you achieve your weight goal. Include type(s) of activity, as well as frequency, duration, and intensity. Cite page numbers in the text to support your answer. Use a separate piece of paper to write your response.

Evaluation of a Commercial Weight-Loss Program

1. What is the name of the weight-loss program you will evaluate?

____

2. Describe the food program promoted by the organization. Is the program flexible enough to allow people with different foodways and lifestyles to use it successfully? Is it adaptable and easy to follow? Does it provide variety? Does the program provide for weight maintenance after goal weight is achieved?

____

____

3. Describe how the plan works. Does it require substantial registration fees, mandatory purchase of various items or foods, attendance at a minimum number of meetings?

____

____

4. Evaluate whether the plan offers a nutritionally sound way to lose weight. What characteristics make it sound or unsound?

____

____

5. Describe any "case histories" you might overhear or learn about and your impressions of them. Do the stories sound authentic? Do they sound as though the person's weight loss was achieved sensibly? Do they present facts? Do they contain any "magic bullets"?

____

____

6. Describe your overall impressions of the meeting, telling whether they were positive or negative and why. If you had a weight problem, would you consider membership in this organization?

____

____

PRACTICE TEST

The following items will help you evaluate your understanding of this lesson. Use the answer key at the end of the lesson to check your answers or to locate material related in each question.

Multiple Choice

Select the one choice that best answers the question.

1. All of the following describe the behavior of fat cells EXCEPT __________.
   A. the number decreases when fat is lost from the body
   B. the storage capacity for fat depends on both cell number and cell size
   C. the number increases at a faster rate in obese children than in lean children
   D. the number increases several-fold during the growth years and tapers off when reaching adult status

2. The theory that the body tends to maintain a certain weight by means of its own internal controls is referred to as the __________.
   A. enzyme theory
   B. fat cell theory
   C. thermogenesis theory
   D. set-point theory

3. What is the incidence of obesity in a person whose parents are both obese?
   A. 10 percent
   B. 30 percent
   C. 90 percent
   D. 98 precent

4. Television watching contributes to obesity for all of the following reasons EXCEPT __________.
   A. it promotes inactivity
   B. it promotes between-meal snacking
   C. it replaces time that could be spent eating
   D. it gives high exposure to energy-dense foods featured in the commercial advertisements

5. Susie has been overweight for many years. She tends to eat when she is not really hungry and eats more food when she is angry or depressed. Which of the following theories could explain Susie's problem of overweight?
   A. Enzyme theory
   B. Set-point theory
   C. External cue theory
   D. Fat cell theory

6. The obesity treatment drugs PPA were removed from the market in 1997 due to case reports of __________.
   A. cataracts
   B. liver failure
   C. kidney failure
   D. pulmonary hypertension

7. A person trying to gain weight should __________.
   A. participate in physical activity
   B. eat more frequently
   C. eat faster
   D. A and B

8. Which of the following would NOT be a recommendation of weight-reduction counseling?
   A. Reorganize some established behavior patterns
   B. Use the exchange system to keep track of energy intake
   C. Perform physical exercise to increase energy expenditure
   D. Reduce daily energy intake to less than 1,000 calories to overcome the decrease in metabolic rate

9. Diet recommendations for people with bulimia include all of the following EXCEPT __________.
   A. eat sitting down
   B. include fiber-rich foods
   C. eat cold foods to stimulate satiety
   D. avoid "finger" foods to minimize overeating

10. Compared to average weight for women in the United States, fashion models are __________.
    A. 10 percent below the average
    B. 18 percent below the average
    C. 23 percent below the average
    D. 35 percent below the average

11. If people need to lose 50 pounds of weight and only achieve a 25-pound weight loss, the criteria established by the medical profession consider those people to be __________.
    A. underachievers
    B. failures
    C. successful
    D. hopeless

12. People with high waist-to-hip ratios have a higher risk for heart disease and are characterized by shape as __________.
    A. "apples"
    B. "slugs"
    C. "pears"
    D. "sloths"

13. A cycle of quick weight loss and weight regain which causes the body's metabolism to drop is known as __________.
    A. cyclic regaining
    B. dieting madness
    C. yo-yo dieting
    D. on-off cycling

14. An extreme surgical procedure, such as a gastric bypass, produces dramatic weight loss in morbidly obese people by __________.
    A. restricting the size of the stomach so it holds less food
    B. stimulating the hypothalamus to not respond to excess food
    C. shortening the GI tract so that food is not absorbed
    D. reducing the length of the colon so food passes quickly

15. Of the most common eating disorders, the one which is most medically dangerous is __________.
    A. bulimia nervosa
    B. anorexia nervosa
    C. obesity nervosa
    D. purging nervosa

16. All of the following are acceptable treatments for bulimia EXCEPT ________.
    A. helping the person normalize eating patterns
    B. focusing on the person's social world
    C. treating all people with bulimia in a hospital setting
    D. changing the attitudes regarding weight and shape

17. Experts believe that the most sound advice for losing weight is __________.
    A. exercise more and eat a healthful diet
    B. balance the calories in with the calories out
    C. eat less than 1,000 calories per day
    D. all of the above

18. The best advice regarding body weight and pregnancy is __________.
    A. lose weight as fast as possible after pregnancy
    B. don't gain any weight during pregnancy
    C. lose weight slowly to keep it off permanently
    D. eat a lot, exercise a lot, then lose a lot

Matching

Select the letter next to the word or phrase that best matches the numbered statements. Use answers only once.

_____	19. An enzyme that promotes fat storage	A. Liposuction
_____	20. A surgical procedure for treating severe obesity	B. Gastric bypass
_____	21. Cause of an increase of basal metabolic rate	C. Lipoprotein lipase
		D. Exercise

Essay Questions

Reflect on the items listed below; one or more may be included on the exam.

22. List the major causes of obesity. Which ones can be controlled by dietary manipulations or behavior modification?

23. List several factors that help identify inappropriate, unsound, and possibly dangerous commercial weight-loss programs.

ANSWER KEY

The following provides the answers and references for the practice test questions.

	Answer	Learning Objectives	References
1.	A	LO 1	Sizer/Whitney, p. 326
2.	D	LO 1	Sizer/Whitney, p. 326
3.	C	LO 1	Sizer/Whitney, p. 327
4.	C	LO 1	Sizer/Whitney, p. 327
5.	C	LO 1	Sizer/Whitney, pp. 326–327
6.	D	LO 3	Sizer/Whitney, p. 345
7.	D	LO 5	Sizer/Whitney, pp. 343–344
8.	D	LO 4	Sizer/Whitney, pp. 336–337
9.	C	LO 6	Sizer/Whitney, pp. 355–356
10.	C	LO 7	Video
11.	C	LO 2	Video
12.	A	LO 8	Video
13.	C	LO 3	Video
14.	A	LO 3	Video
15.	B	LO 6	Video
16.	C	LO 6	Video
17.	A	LO 4	Video
18.	C	LO 4	Video
19.	C	LO 1	Sizer/Whitney, p. 326
20.	B	LO 3	Sizer/Whitney, p. 346
21.	D	LO 4	Sizer/Whitney, p. 342
22.		LO 1	Sizer/Whitney, pp. 325–328
23.		LO 2, 3	Sizer/Whitney, pp. 344–346

Lesson 12

Vitamins: Water Soluble

OVERVIEW

Some nutrients are considered non-energy-yielding nutrients—the vitamins and minerals. This lesson begins with vitamins B and C—the water-soluble vitamins—and examines their roles in the body, the deficiency and/or toxicity associated with their intake, and their major food sources. You will also see how B vitamins work in concert to aid the body in energy production, how vitamin C works its antioxidant effects on the body, and how vitamins can be increased for athletes.

LEARNING OBJECTIVES

Upon completing this lesson, you should be able to:

1. Distinguish among vitamins and the energy-yielding nutrients.
2. Describe in general terms how B vitamins work together in the body.
3. Identify the function(s) in the body associated with each water-soluble vitamin.
4. Discuss the deficiency symptoms and/or disease (if any) of each water-soluble vitamin.
5. Discuss toxicity symptoms and/or disease (if any) of each water-soluble vitamin.
6. Identify major food sources for each water-soluble vitamin.
7. Identify the factors that destroy each of the water-soluble vitamins and explain how vitamin losses can be prevented during handling, storage, and preparation.
8. Explain the antioxidant effects of vitamin C on the body.

LESSON ASSIGNMENT

Review the following assignment in order to schedule your time appropriately. Pay careful attention; the titles and numbers of the textbook chapter, the telecourse guide lesson, and the video program may be different from one another.

Text: Sizer and Whitney, *Nutrition: Concepts and Controversies*, Chapter 7, "The Vitamins," p. 213–262

Video: "Vitamins: Water Soluble" from the series *Nutrition Pathways*

LESSON FOCUS POINTS

The following questions are designed to help you get the most out of the lesson. After you read the textbook assignments and view the video, you should be able to answer these questions.

1. How do vitamins differ from energy-yielding nutrients with regard to structure, function, and amounts found in foods? Why should one eat water-soluble vitamins more often than fat-soluble vitamins? (text, p. 214)
2. Define *antioxidant*. How does vitamin C act as an antioxidant? (text, pp. 221–229)
3. How might vitamin C affect the immune system? (text, p. 229; video)
4. What is the disease caused by vitamin C deficiency? How does vitamin C function with regard to collagen formation? Why does cigarette smoking increase the need for vitamin C? What are the earliest signs and symptoms of vitamin C deficiency? What are the symptoms associated with vitamin C toxicity? What are the foods which provide generous amounts of vitamin C? Does taking high doses of vitamin C prevent colds? (text, pp. 229–232)
5. How do B vitamins interact with energy-yielding foods, especially carbohydrate and fat? (text, p. 233; video)
6. How do the B vitamins function in general in the body? How do B vitamins act as coenzymes? (text, pp. 233–235)

7. What disease is associated with thiamin deficiency? What foods are exceptional sources of thiamin? (text, pp. 235–236)

8. What systems in the body are affected by riboflavin deficiency? What is the collective name given to riboflavin deficiencies? What is the top food source for riboflavin? (text, pp. 236–237)

9. What role does niacin play in metabolism? Name the amino acid that can be used in the body to make niacin. What disease is associated with niacin deficiency? What is the "niacin flush"? What disease or medical condition is sometimes treated with niacin? What types of foods are best for niacin equivalents in the diet? (text, pp. 236–238)

10. What is the main function of folate? With what other B vitamin does folate interact? Why are folate recommendations increased for pregnant women? How does folate deficiency affect the body? What foods have abundant amounts of folate? (text, pp. 238–240)

11. Describe the relationship between B_{12} and folate. What is an "intrinsic factor"? What other B vitamin deficiency results from a B_{12} deficiency? What is the most likely reason for developing a B_{12} deficiency? In what foods is B_{12} found almost exclusively? How does this impact strict vegetarians? (text, pp. 240–241)

12. How does B_6 affect amino acid metabolism? What are the symptoms of B_6 deficiency? How did researchers discover that B_6 had toxic levels? What are the toxic effects of B_6? What are the best food sources for B_6? (text, pp. 241–243)

13. How does biotin generally function in energy production? What are the sources of biotin for humans? What is the consequence of eating large amounts of raw eggs? (text, pp. 243–244)

14. What are particularly good food sources of pantothenic acid? (text, pp. 243–244)

15. What are suggestions to prevent vitamin loss during handling, storage, and preparation of vegetables for home use and for extended periods? (video)

KEY CONCEPTS

Factors that Destroy Vitamins

Nutrient	Acid	Alkaline	Oxygen	UV Light	Heat
B Vitamins:					
Thiamin		X	X		X
Riboflavin		X		X	
Niacin	X				
Biotin	X	X	X	X	
Pantothenic acid	X	X			X
Vitamin B_6		X		X	X
Folate	X		X	X	X
Vitamin B_{12}	X			X	
Vitamin C		X	X		X
Vitamin A			X	X	X
Vitamin D	X		X	X	
Vitamin E			X	X	
Vitamin K		X		X	

Source: Adapted from C. D. Berdanier, Advanced Nutrition: Micronutrients (Boca Raton, FL: CRC Press, 1998).

RELATED ACTIVITIES

This activity is not required unless your instructor assigns it. It is offered as a suggestion to help you learn more about the material presented in this lesson.

Complete the form on the following page and return it to your instructor according to the established deadline.

❑ Fill in the Table of Water-Soluble Vitamins

Fill in the Table of Water-Soluble Vitamins

In the space provided, write in three of each of the following: the major functions, deficiency disease and/or symptoms, toxicity symptoms, and the major food sources for each of the B vitamins and vitamin C. If a characteristic does not apply, write NA in the space.

Vitamin (Coenzyme)	Function(s) of Vitamin	Deficiency Disease and/or Symptoms	Toxicity Symptoms	Major Food Sources
B_1, Thiamin	1. 2. 3.	1. 2. 3.	1. 2. 3.	1. 2. 3.
B_2, Riboflavin	1. 2. 3.	1. 2. 3.	1. 2. 3.	1. 2. 3.
B_3, Niacin	1. 2. 3.	1. 2. 3.	1. 2. 3.	1. 2. 3.
Folate, Folic Acid, Folacin	1. 2. 3.	1. 2. 3.	1. 2. 3.	1. 2. 3.
B_{12}, Cobalamin	1. 2. 3.	1. 2. 3.	1. 2. 3.	1. 2. 3.
B_6	1. 2. 3.	1. 2. 3.	1. 2. 3.	1. 2. 3.
Biotin	1. 2. 3.	1. 2. 3.	1. 2. 3.	1. 2. 3.
Pantothenic Acid	1. 2. 3.	1. 2. 3.	1. 2. 3.	1. 2. 3.
C, Ascorbic Acid	1. 2. 3.	1. 2. 3.	1. 2. 3.	1. 2. 3.

PRACTICE TEST

The following items will help you evaluate your understanding of this lesson. Use the answer key at the end of the lesson to check your answers or to locate material related in each question.

Multiple Choice

Select the one choice that best answers the question.

1. What is the primary excretory route for the water-soluble vitamins?
   A. Bile
   B. Kidney
   C. Intestine
   D. Perspiration

2. The B vitamins act as part of __________.
   A. anticoagulants
   B. antibodies
   C. coenzymes
   D. intrinsic factors

3. The disease caused by thiamin deficiency is known as __________.
   A. pellagra
   B. osteomalacia
   C. scurvy
   D. beriberi

4. Of the following, which is the richest food source of thiamin?
   A. Milk
   B. Pork
   C. Lettuce
   D. Refined rice

5. Pellagra is a disease caused by a deficiency of __________.
   A. vitamin C
   B. niacin
   C. thiamin
   D. vitamin B_{12}

6. Which of the following is a feature of niacin nutrition?
   A. Low doses may lead to kidney stones
   B. High doses may lower blood lipids
   C. Low doses may lead to heartburn and low blood pressure
   D. High doses may elevate red blood cell count in mildly anemic individuals

7. Which of the following activities is shared by vitamin B_{12} and folate?
   A. Both are required in the making of red blood cells
   B. Both require intrinsic factor for their release from food proteins
   C. Both are found in significant amounts in green leafy vegetables
   D. Both are considered problem nutrients for strict vegetarians

8. Which of the following B vitamins is especially important for women of childbearing age, to prevent neural tube defects?
   A. Vitamin B_6
   B. Biotin
   C. Folate
   D. Vitamin B_{12}

9. Which of the following represents the results of well-controlled studies of vitamin C supplementation on the resistance to, and recovery from, colds?
   A. There was a reduction in the duration of colds by 50 percent on the average
   B. There was only a minor effect on reducing the number and severity of colds
   C. There was a significant reduction in the duration of colds only in people who consumed at least 1 gram a day
   D. There was a significant reduction in the number of colds only in people who consumed more than 3 grams per day

10. Which of the following vitamins is known to deactivate histamine, a substance that causes nasal congestion?
    A. Niacin
    B. Vitamin E
    C. Vitamin C
    D. Vitamin B_{12}

11. In which of the following ways do B vitamins interact with energy-yielding foods?
   A. They provide an essential source of additional energy.
   B. They are not a source of energy, but they are needed to change the foods we eat into energy.
   C. B vitamins only interact with fatty acids to produce energy.
   D. B vitamins only interact with carbohydrates to produce energy.

12. As far as preparing vegetables is concerned, the key in preventing vitamin losses is to __________.
   A. cook until soft to the touch
   B. not overcook—keep slightly crisp
   C. use a lot of water when cooking
   D. drink the water in which the vegetables were cooked

13. When studying allergies and aging, research has found a positive effect from vitamin ____.
   A. C
   B. B_6
   C. E
   D. B_{12}

Matching

Select the letter next to the word or phrase that best matches the numbered statements. Use answers only once.

_____ 14. Used for synthesis of niacin

_____ 15. Toxicity from this vitamin is known to cause irreversible nerve damage

_____ 16. One of the first symptoms of folate deficiency

_____ 17. Vitamin C is required for the synthesis of this substance

A. Anemia
B. Collagen
C. Tryptophan
D. Vitamin B_6
E. Vitamin B_{12}

Essay Questions

Reflect on the items listed below; one or more may be included on the exam.

18. Under what circumstances can water-soluble vitamins be toxic? Cite several examples.

19. Discuss the roles of the B vitamins in energy metabolism.

20. Under what conditions and for what reasons would intakes of vitamin C above the RDA be desirable?

ANSWER KEY

The following provides the answers and references for the practice test questions.

	Answer	Learning Objectives	References
1.	B	LO 5	Sizer/Whitney, p. 228
2.	C	LO 2	Sizer/Whitney, p. 233
3.	D	LO 4	Sizer/Whitney, p. 235
4.	B	LO 6	Sizer/Whitney, p. 235
5.	B	LO 4	Sizer/Whitney, p. 236
6.	B	LO 3	Sizer/Whitney, pp. 237
7.	A	LO 3	Sizer/Whitney, p. 240
8.	C	LO 3	Sizer/Whitney, pp. 238
9.	B	LO 3	Sizer/Whitney, p. 231
10.	C	LO 8	Sizer/Whitney, p. 231
11.	B	LO 1	Video
12.	B	LO 7	Video
13.	A	LO 8	Video
14.	C	LO 6	Sizer/Whitney, p. 237
15.	D	LO 5	Sizer/Whitney, p. 243
16.	A	LO 4	Sizer/Whitney, p. 240
17.	B	LO 3	Sizer/Whitney, p. 229
18.		LO 5	Sizer/Whitney, pp. 224, 230, 234, 236, and 247–249
19.		LO 1, 2	Sizer/Whitney, pp. 233–235
20.		LO 8	Sizer/Whitney, pp. 230–231

Lesson 13

Vitamins: Fat Soluble

OVERVIEW

In this lesson, you will learn that there are many differences between the fat-soluble and water-soluble vitamins. Fat-soluble vitamins are found in the oils and fats of foods, are insoluble in water, need bile to be digested, require chylomicrons for transport, go through the lymphatic system before entering the bloodstream, and are stored in the liver and fat tissues. You will learn what happens when you don't have enough of the fat-soluble vitamins in your diet. Because they are not readily excreted, you will also see how taking excessive fat-soluble vitamins can have toxic effects and why we only need to take them over a period of several days rather than daily. Finally, you will recognize the function of vitamin E as an antioxidant.

LEARNING OBJECTIVES

Upon completing this lesson, you should be able to:

1. Explain the differences between water- and fat-soluble vitamins.
2. Identify the primary function(s) in the body of each fat-soluble vitamin.
3. Discuss the deficiency symptoms and/or disease (if any) of each fat-soluble vitamin.
4. Discuss the toxicity symptoms and/or disease (if any) of each fat-soluble vitamin.
5. Identify major food sources for each fat-soluble vitamin.
6. Explain how vitamin E acts as an antioxidant.
7. Distinguish between the advantages and disadvantages of vitamin/mineral supplementation.
8. Discuss guidelines for choosing an appropriate supplement.

LESSON ASSIGNMENT

Review the following assignment in order to schedule your time appropriately. Pay careful attention; the titles and numbers of the textbook chapter, the telecourse guide lesson, and the video program may be different from one another.

Text: Sizer and Whitney, *Nutrition: Concepts and Controversies*, Chapter 7, "The Vitamins," pp. 213–262

Video: "Vitamins: Fat Soluble" from the series *Nutrition Pathways*

LESSON FOCUS POINTS

The following questions are designed to help you get the most out of the lesson. After you read the textbook assignments and view the video, you should be able to answer these questions.

1. Describe how fat-soluble vitamins differ from water-soluble vitamins for the following: solubility, absorption/transport, storage, excretion, frequency of intake, and risk for toxicity. (text, pp. 214–215, 228)

2. Describe the two roles of vitamin A in the eyes. How does vitamin A function in bone remodeling? What causes vitamin A toxicity? What is the recommendation for limiting vitamin A supplements for pregnant women and why? What is the recommendation regarding vitamin A and acne treatments? What foods are the best source of preformed vitamin A? (text, pp. 215–219)

3. How can vitamin A toxicity affect the brain? (text, p. 218; video)

4. How does beta-carotene act as an antioxidant? What can happen if a person eats too many foods containing beta-carotene? What foods contain carotenoids? What carotenoid has the highest vitamin A activity? What color in vegetables is typically associated with Vitamin A content? (text, pp. 220–221)

5. How is cholesterol associated with vitamin D? What organs are the primary targets of vitamin D activity? What is the special function of vitamin D in bone growth? What is rickets? What is osteomalacia? What are the symptoms of vitamin D toxicity? How does vitamin D compare to other fat-soluble vitamins with regard to toxic effects? What are the best food sources of

vitamin D? How much time in the sun is needed to maintain vitamin D status? (text, pp. 221–224)

6. Explain the role of vitamin E as an antioxidant. (text, p. 224)

7. What is the classic sign of vitamin E deficiency? How does vitamin E toxicity compare to toxicity associated with other fat-soluble vitamins? What can happen with extremely high doses of vitamin E? What are the best food sources of vitamin E? (text, pp. 224–226)

8. What is the primary function of vitamin K? What nonfood source contributes to vitamin K levels? Under what two circumstances can vitamin K deficiency develop? When can vitamin K toxicity occur? What are the best food sources of vitamin K? (text, pp. 227–228)

9. Who needs to take vitamin/mineral supplements? Which supplements are helpful and which should be avoided? Are dietary supplements safe? (text, pp. 255–262)

10. What are two basic questions to ask when choosing a supplement? What level of vitamin and minerals does an appropriate supplement contain? What should be avoided when choosing a supplement? (text, pp. 255–262)

11. Under what circumstances might a doctor prescribe antioxidation supplementation? What is vitamin E's chief role as an antioxidant? What diseases might be prevented through the use of vitamins E and C? (text, p. 260; video)

12. What is the advice regarding antioxidant supplementation, exercise, and immunity? (video)

KEY CONCEPTS

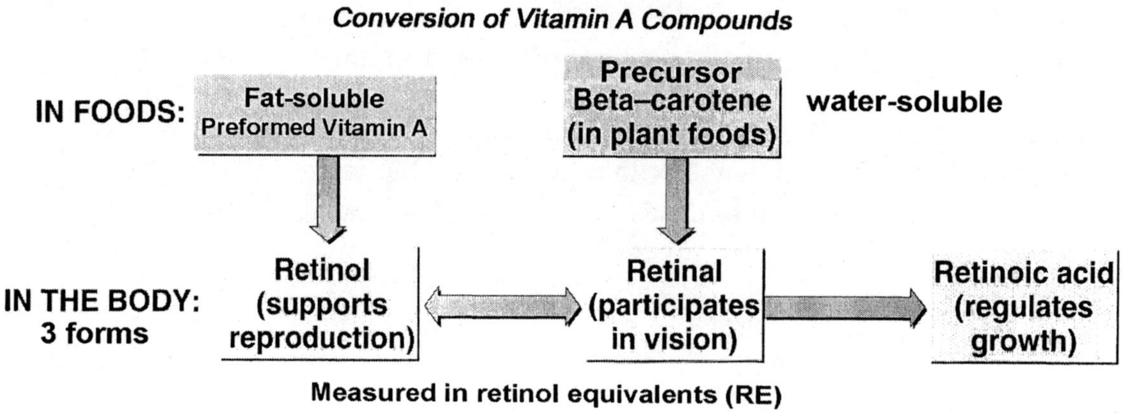

© 1999 Wadsworth Publishing Company/ITP

Classifications of Vitamins

Characteristics	Water Soluble C, B Vitamins	Fat Soluble A, E, D, K
Absorption	Blood	Lymph → blood
Transport	Travel freely	Protein carrier
Storage	Freely circulate in water in body	With fat in the cell
Excretion	Urinary excretion	Remain in fat storage site
Toxicity	Excess not likely to be toxic	Excess can be toxic
Requirement	Frequent, small doses	Periodic doses

The Theory of Antioxidants and Diseases

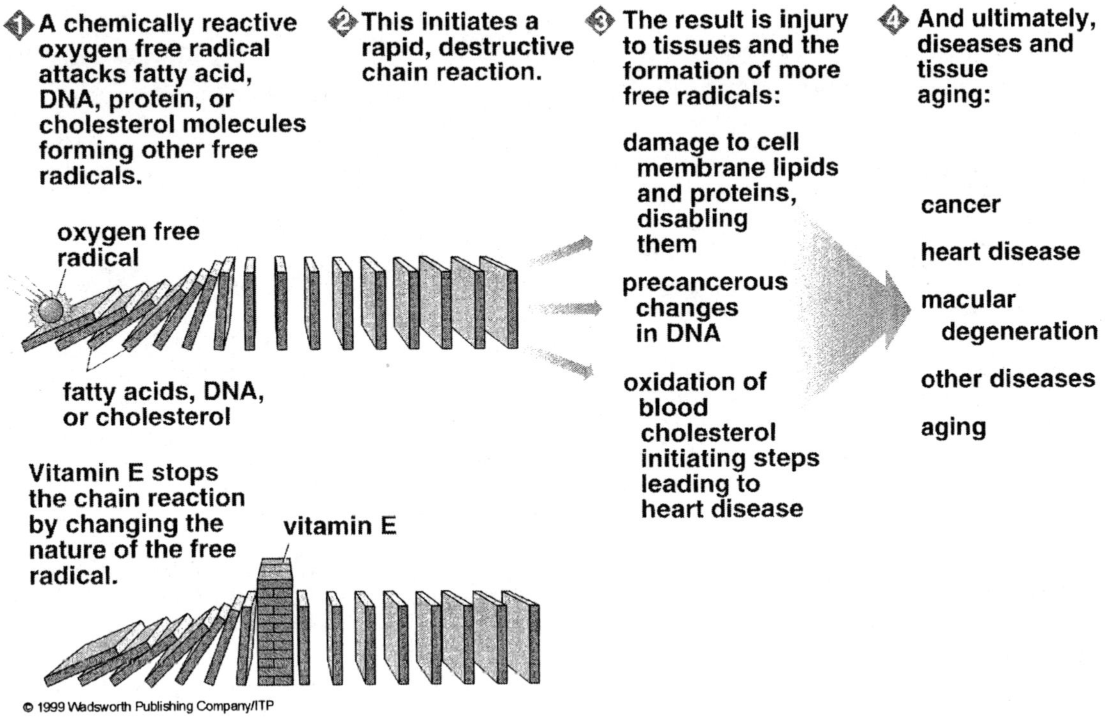

RELATED ACTIVITIES

These activities are not required unless your instructor assigns them. They are offered as suggestions to help you learn more about the material presented in this lesson.

Complete the following forms and return them to your instructor according to the established deadline.

- ❑ Fill in the Table of Fat-Soluble Vitamins
- ❑ Vitamin/Mineral Supplement Evaluation

Fill in the Table of Fat-Soluble Vitamins

In the space provided, write in three of each of the following: the major functions, deficiency disease and/or symptoms, toxicity symptoms, and the major food sources for each of the fat-soluble vitamins. If a characteristic does not apply, write NA in the space.

NOTE: Vitamin A and beta-carotene are treated separately.

Name of Vitamin	Function(s) of Vitamin	Deficiency Disease and/or Symptoms	Toxicity Symptoms	Major Food Sources
A (Retinol)	1. 2. 3.	1. 2. 3.	1. 2. 3.	1. 2. 3.
Beta-carotene	1. 2. 3.	1. 2. 3.	1. 2. 3.	1. 2. 3.
D	1. 2. 3.	1. 2. 3.	1. 2. 3.	1. 2. 3.
E (D-alpha-tocopherol)	1. 2. 3.	1. 2. 3.	1. 2. 3.	1. 2. 3.
K	1. 2. 3.	1. 2. 3.	1. 2. 3.	1. 2. 3.

Vitamin/Mineral Supplement Evaluation

Complete the supplement evaluation form about a selected supplement.

1. What is the name of the supplement? ________________________________

2. What is the cost per pill? ____

3. Is the supplement complete? (Does it contain all vitamins and minerals with established reference daily intakes [RDIs]?) If no, what is missing?

   ____

   ____

4. Are most vitamins and minerals present at or near 100 percent of the RDIs? Exceptions include biotin, calcium, magnesium, and phosphorus, which are rarely found in amounts near 100 percent of the RDIs. List any vitamins or minerals that are present in low amounts or dangerously high amounts.

   ____

   ____

5. Does the supplement contain unnecessary nutrients or nonnutrients? If yes, list them.

   ____

   ____

6. Is there "hype" on the label? Does the label use the terms *natural*, *organic*, *chelated*, *no sugar*, *stress-reliever*, etc? List any terms used.

   ____

   ____

PRACTICE TEST

The following items will help you evaluate your understanding of this lesson. Use the answer key at the end of the lesson to check your answers or to locate material related in each question.

Multiple Choice

Select the one choice that best answers the question.

1. True statements concerning the water-soluble vitamins include:
   A. They are stored in body tissues to a large extent.
   B. They are easily excreted in the urine.
   C. Pose a greater risk for toxicity than fat-soluble vitamins.
   D. A and B.

2. Vitamin A plays an important role in __________.
   A. maintenance of body linings
   B. blood clotting
   C. maintenance of a healthy immune system
   D. A and C

3. Vitamin A supplements are helpful in treating which of the following conditions?
   A. Acne
   B. Rickets
   C. Osteomalacia
   D. Night blindness

4. Which of the following provides the LEAST amount of beta carotene?
   A. Corn
   B. Spinach
   C. Carrots
   D. Cantaloupe

5. In what system would the effects of a vitamin D deficiency be most readily observed?
   A. Nervous
   B. Skeletal
   C. Muscular
   D. Circulatory

6. The main function of vitamin E in the body is to act as __________.
   A. a coenzyme
   B. a precursor
   C. an antioxidant
   D. a free radical

7. Among the following, which contains the highest concentration of vitamin E?
   A. Butter
   B. Carrots
   C. Milk fat
   D. Corn oil

8. A nonfood source from which vitamin K can be obtained is __________.
   A. sunlight
   B. intestinal bacteria
   C. antibiotics
   D. A and B

9. What is a free radical?
   A. A highly reactive, unstable molecule because it contains unpaired electrons
   B. An antioxidant substance that prevents accumulation of cell-damaging oxides
   C. A substance in food that interacts with nutrients to decrease their utilization
   D. A nutrient in excess of body needs that the body is free to degrade with no consequence

10. If taken in excess, vitamin A can produce adverse effects, such as __________.
   A. pressure in the brain which can cause headaches
   B. blurred vision which can cause headaches
   C. bone mineralization which can cause rickets
   D. all of the above

11. Rust forming on a metal surface or butter turning rancid are examples of __________.
   A. an oxidation reaction which produces undesirable results
   B. a mineral interacting with a hard surface
   C. iron particles interacting with water molecules
   D. B and C

12. Several conditions and/or diseases such as some cancers, aging, and heart disease might be prevented in the future through the use of __________.
   A. antioxidant supplementation with vitamins C and E
   B. medications that have yet to be discovered
   C. new exercise equipment that is currently being developed
   D. all of the above

13. Young patients who have heart disease who do not have risk factors that can be treated might be advised to take __________.
   A. vitamin A supplements
   B. vitamin E supplements
   C. vitamin D supplements
   D. multiple vitamins

14. Vitamin E supplements are often taken by runners to help prevent damage to lungs caused by __________.
   A. heavy exercise during cold weather
   B. exercising when air pollution is high
   C. exercising when they have a respiratory infection
   D. all of the above

Fill in the Blank

Insert the correct word or words in the blank for each item.

15. When rust forms on an iron railing or cooking oil smells rancid, the undesirable effect is caused by a process known as ____________________.

16. The body system that appears to improve when elderly people take vitamin E and other antioxidants is ___________________________.

Matching

Select the letter next to the word or phrase that best matches the numbered statements. Use answers only once.

______	17. Pigment in carrots and pumpkins	A. Carotene
______	18. Another term for blood clotting	B. Coagulation
______	19. Food source of vitamin K	C. Green leafy vegetables
		D. Keratin

Essay Questions

Reflect on the items listed below; one or more may be included on the exam.

20. Describe how the body can synthesize active vitamin D with the help of sunlight.

21. Discuss the pros and cons of taking antioxidant supplements.

ANSWER KEY

The following provides the answers and references for the practice test questions.

	Answer	Learning Objectives	References
1.	B	LO 1	Sizer/Whitney, pp. 214–215
2.	D	LO 2	Sizer/Whitney, p. 215
3.	D	LO 3	Sizer/Whitney, p. 216
4.	A	LO 5	Sizer/Whitney, pp. 219–220
5.	B	LO 3	Sizer/Whitney, p. 221
6.	C	LO 3	Sizer/Whitney, p. 224
7.	D	LO 5	Sizer/Whitney, p. 226
8.	B	LO 5	Sizer/Whitney, p. 227
9.	A	LO 6	Sizer/Whitney, p. 224
10.	A	LO 4	Video
11.	A	LO 6	Video
12.	A	LO 6	Video
13.	B	LO 7, 8	Video
14.	B	LO 6, 8	Video
15.	oxidation	LO 6	Video
16.	the immune system (immunity, immune response)	LO 6	Video
17.	A	LO 5	Sizer/Whitney, p. 220
18.	B	LO 2	Sizer/Whitney, p. 227
19.	C	LO 5	Sizer/Whitney, p. 228
20.		LO 5	Sizer/Whitney, pp. 222–223
21.		LO 6, 7	Sizer/Whitney, pp. 258–260

Lesson 14

Major Minerals and Water

OVERVIEW

"Drink eight glasses of water a day." You've heard this before, so you know that water must be good for you. You could live for weeks, even months, without food, but you would not live long, maybe a few days, without water—it's that important to life. In this lesson, you will learn how critical water is to normal, healthy functioning. In addition, you will examine other nutrients that are necessary for good health—the major minerals. You will discover that minerals are inorganic compounds that appear in our bodies in the same form as they appear in nature—indestructible—and essential to life. Further, you will also learn what deficiencies can develop when both minerals and water are not consumed in adequate amounts. Finally, you will focus attention on the development, symptoms, and treatment of osteoporosis, a condition directly related to the major mineral calcium.

LEARNING OBJECTIVES

Upon completing this lesson, you should be able to:

1. Discuss the major roles of water in the body.
2. Describe how the body regulates and maintains water balance.
3. Cite the recommendations for water intake for adults.
4. Briefly explain the role of minerals in maintaining fluid and electrolyte balance.
5. Identify the chief functions of the major minerals in the body.
6. Discuss the deficiency symptoms and/or disease for the major minerals.
7. Discuss the toxicity symptoms and/or disease for the major minerals.
8. List food sources for the major minerals.

9. Cite the minimum recommended intake and the current health recommendation for sodium intake.

10. Describe the development, symptoms, and treatment of osteoporosis.

LESSON ASSIGNMENT

Review the following assignment in order to schedule your time appropriately. Pay careful attention; the titles and numbers of the textbook chapter, the telecourse guide lesson, and the video program may be different from one another.

Text: Sizer and Whitney, *Nutrition: Concepts and Controversies*, Chapter 8, "Water and Minerals," pp. 263–310

Video: "Major Minerals and Water" from the series *Nutrition Pathways*

LESSON FOCUS POINTS

The following questions are designed to help you get the most out of the lesson. After you read the textbook assignments and view the video, you should be able to answer these questions.

1. Cite the functions of water in the body. (text, pp. 265–266)

2. How much of an adult's body consists of water? What factors influence water intake? How do the mouth and hypothalamus help maintain water balance? What are the sources of water for the body? How much water per day does a person derive from these sources? How much water per day must the body excrete to rid itself of waste products? How much water per day does a person need under normal environmental circumstances? What is dehydration? What is water intoxication? (text, pp. 265–268)

3. What is an ion? What is an electrolyte? What is fluid and electrolyte balance? What are circumstances which could cause electrolyte/fluid imbalances? Which minerals are the first to be lost through fluid losses? How do minerals help in maintaining acid-base balance? (text, pp. 272–273)

4. Where is calcium found in the body? What two roles does calcium play in the bones? What is hydroxyapatite? How does calcium function in body fluids? What body systems help raise blood calcium when it falls too low? What factors help increase calcium absorption? What are the recommendations for calcium intake? What food provides the single most abundant source of calcium? What are other food sources of available calcium? What deficiency develops with inadequate calcium intake? (text, pp. 273–276)

5. Where is phosphorus found in the body? How does phosphorus interact with calcium? Describe the roles phosphorus plays in the body. What are the best sources of phosphorus in the diet? (text, pp. 276–277)

6. Where is magnesium found in the body? How does magnesium function in the body? What foods provide the most magnesium to the diet? What conditions might produce magnesium deficiency? (text, pp. 277–278)

7. How does sodium function in the body? What is the recommended minimum sodium intake per day? What is the health recommendation regarding daily salt intake? What is the connection between sodium and hypertension? Which foods provide the most sodium to the diet? (text, pp. 279–281)

8. What are the roles of potassium in the body? What foods provide abundant amounts of potassium? What conditions can produce a potassium deficiency? What are the toxic effects of potassium? (text, pp. 281–283)

9. What are the functions of chloride in the body? What is the principal food source of chloride? (text, pp. 283–284)

10. What role does sulfur play in the body? What foods contain sulfur? (text, p. 284)

11. What factors increase the risk for developing osteoporosis? (text, pp. 303–310)

12. How much body weight in the form of water could be lost during an hour's worth of exercise? What percent of water loss could result in a life-threatening situation? What is the very first sign of dehydration? What physical symptoms might be displayed by a child who is dehydrated? What advice is given to parents of a dehydrated child? (video)

13. What amount of sodium per day is needed for an average adult? (text, p. 279; video)

14. How is osteoporosis characterized? Which population is most affected? What are the most common causes of osteoporosis? What are the risk factors associated with osteoporosis? What is the treatment for osteoporosis with regard to the following: exercise, calcium, salt, and animal products? What is a recent, novel approach to treatment and what have been the results of the treatment? How can osteoporosis be prevented? What is often the first symptom of osteoporosis? (text, pp. 303–310; video)

15. How can increased water intake impact sodium's adverse effects on the body? (video)

RELATED ACTIVITY

This activity is not required unless your instructor assigns it. It is offered as a suggestion to help you learn more about the material presented in this lesson.

Complete the form on the following page and return it to your instructor according to the established deadline.

❏ Fill in the Table of Major Minerals

Fill in the Table of Major Minerals

In the space provided, list three of the following: the major functions, deficiency disease and/or symptoms, toxicity symptoms, and the major food sources for the major minerals. If a characteristic does not apply, write NA in the space.

Name of Mineral	Function(s) of Mineral	Deficiency Disease and/or Symptoms	Toxicity Symptoms	Major Food Sources
Calcium	1. 2. 3.	1. 2. 3.	1. 2. 3.	1. 2. 3.
Phosphorus	1. 2. 3.	1. 2. 3.	1. 2. 3.	1. 2. 3.
Magnesium	1. 2. 3.	1. 2. 3.	1. 2. 3.	1. 2. 3.
Sodium	1. 2. 3.	1. 2. 3.	1. 2. 3.	1. 2. 3.
Potassium	1. 2. 3.	1. 2. 3.	1. 2. 3.	1. 2. 3.
Chloride	1. 2. 3.	1. 2. 3.	1. 2. 3.	1. 2. 3.
Sulfur	1. 2. 3.	1. 2. 3.	1. 2. 3.	1. 2. 3.

PRACTICE TEST

The following items will help you evaluate your understanding of this lesson. Use the answer key at the end of the lesson to check your answers or to locate material related in each question.

Multiple Choice

Select the one choice that best answers the question.

1. Which of the following makes up about 60 percent of the body's weight?
   A. Major minerals
   B. Protein
   C. Water
   D. Trace minerals

2. What organ provides the major control of fluid balance in the body?
   A. Liver
   B. Heart
   C. Kidneys
   D. Skeletal muscle

3. The chief ion used to maintain the volume of fluid outside cells is ________.
   A. potassium
   B. chloride
   C. sodium
   D. calcium

4. How much water should a person who expends about 3,000 calories a day consume?
   A. 2 to 3 liters
   B. 3 to 4½ liters
   C. 4 to 5 liters
   D. 5½ to 7 liters

5. Over half of the body's magnesium is stored in the __________.
   A. teeth
   B. liver
   C. bones
   D. kidneys

6. Almost all (99 percent) of the calcium in the body is used to __________.
   A. provide energy for cells
   B. provide rigidity for the bones and teeth
   C. regulate the transmission of nerve impulses
   D. maintain the blood level of calcium within very narrow limits

7. The best way to obtain the calcium you need is to __________.
   A. consume more spinach, rhubarb, and Swiss chard
   B. eat more cottage cheese and frozen yogurt
   C. consume more milk, cheese, and yogurt
   D. consume more cream cheese, butter, and cream

8. Sulfur is present in practically all __________.
   A. vitamins
   B. proteins
   C. fatty acids
   D. carbohydrates

9. Which of the following statements is true?
   A. People develop their peak bone mass up to the age of thirty.
   B. The skeleton no longer adds to bone density after twenty-four years of age.
   C. The bones begin to lose density after forty years of age regardless of calcium intake.
   D. People do not need calcium throughout life.

10. Among the following ethnic groups, which has the highest bone density?
   A. Japanese
   B. Caucasian
   C. African-American
   D. South American Hispanic

11. It would not be unusual for a person exercising for one hour to lose water weight in the amount of __________.
A. 10 to 15 pounds
B. 5 to 10 pounds
C. 4 to 8 pounds
D. 2 to 4 pounds

12. Some of the physical symptoms reflecting dehydration in a child might include __________.
A. deep, dark circles under the eyes
B. increased heart rate for age
C. dry, tacky mucous membranes in the mouth
D. all of the above

13. Regarding sodium intake, an average adult only needs __________.
A. 5 grams or 1 teaspoon a day
B. 5,000 milligrams or 1 teaspoon a day
C. 500 milligrams or one-tenth of a teaspoon a day
D. A and B

14. The most common cause(s) of osteoporosis __________.
A. is a lack of estrogen in postmenopausal women
B. is increased bone formation in the elderly
C. is lack of weight-bearing exercise during the growing years
D. are all of the above

15. What type of exercise can improve the outcome of osteoporosis?
A. Weight-bearing exercises, such as walking
B. Swimming laps or water aerobics
C. Meditation exercises
D. All of the above

16. An effect of increased water intake on the body of a person who is accustomed to greater quantities of salt is __________.
A. reduced edema or swelling of ankles
B. more bloating
C. more energy
D. all of the above

Fill in the Blank

Insert the correct word or words in the blank for each item.

17. Fatigue is the first reliable sign that a person is ____________________.

18. A slow and steady loss of bone calcium resulting in fracturing or splintering of bones is a characteristic of __________________________.

Essay Questions

Reflect on the items listed below; one or more may be included on the exam.

19. Identify and explain factors involved in governing water intake and water excretion.

20. Define *osteoporosis*. How does it develop? How do age, sex, hormones, and genetics affect its development? How does physical activity affect it? What dietary interventions can help prevent or treat osteoporosis?

ANSWER KEY

The following provides the answers and references for the practice test questions.

	Answer	Learning Objectives	References
1.	C	LO 1	Sizer/Whitney, p. 265
2.	C	LO 2	Sizer/Whitney, p. 267
3.	C	LO 4	Sizer/Whitney, p. 272
4.	B	LO 3	Sizer/Whitney, p. 268
5.	C	LO 5	Sizer/Whitney, p. 277
6.	B	LO 6	Sizer/Whitney, p. 273
7.	C	LO 8	Sizer/Whitney, pp. 275–276
8.	B	LO 8	Sizer/Whitney, p. 284
9.	C	LO 10	Sizer/Whitney, p. 275
10.	C	LO 10	Video
11.	D	LO 2	Video
12.	D	LO 2	Video
13.	C	LO 9	Video
14.	A	LO 10	Video
15.	A	LO 10	Video
16.	A	LO 4, 7	Video
17.	dehydrated	LO 2	Video
18.	osteoporosis	LO 10	Video
19.		LO 2	Sizer/Whitney, pp. 266–267
20.		LO 10	Sizer/Whitney, pp. 274–275

Lesson 15

Trace Minerals

OVERVIEW

Because we need them in such small quantities, the trace minerals are not enough to fill a teaspoon. Without them, however, we would not be able to exist. In this lesson, you will learn about the minor or trace minerals which have an established RDA (iron, zinc, iodine, and selenium) and those which have ranges for safe and adequate daily intakes (copper, fluoride, chromium). You will identify each minor mineral according to the roles it plays in healthy functioning; you will learn what happens when we are deficient in, or consume excessive amounts of, minor minerals; and you will learn what foods provide us with adequate amounts of minor minerals. In addition, you will delve more deeply into the functions of iron throughout pregnancy, adolescence, and in the aging adult as well as exploring how iron-rich foods can be delivered to these populations.

LEARNING OBJECTIVES

Upon completing this lesson, you should be able to:

1. Explain the roles of iron in the body.
2. Describe the effects of iron deficiency.
3. Describe the effects of iron toxicity.
4. Compare and contrast absorption of heme and nonheme iron.
5. List food sources of heme and nonheme iron.
6. Identify the roles of zinc, iodine, selenium, fluoride, chromium, and copper in the body.
7. Discuss symptoms and/or diseases caused by deficiency of zinc, iodine, selenium, fluoride, chromium, and copper.

8. Discuss symptoms and/or diseases resulting from toxicity of zinc, iodine, selenium, fluoride, chromium, and copper.

9. List food sources for zinc, iodine, selenium, fluoride, chromium, and copper.

10. Describe the benefits of phytochemicals in disease prevention to certain populations.

LESSON ASSIGNMENT

Review the following assignment in order to schedule your time appropriately. Pay careful attention; the titles and numbers of the textbook chapter, the telecourse guide lesson, and the video program may be different from one another.

Text: Sizer and Whitney, *Nutrition: Concepts and Controversies*, Chapter 8, "Water and Minerals," pp. 263–310

Video: "Trace Minerals" from the series *Nutrition Pathways*

LESSON FOCUS POINTS

The following questions are designed to help you get the most out of the lesson. After you read the textbook assignments and view the video, you should be able to answer these questions.

1. What is the major role of iodine in the body? What condition develops with an iodine deficiency? What is cretinism? What condition develops with iodine toxicity? What is the major source of the world's iodine? What other foods contribute to iodine intake? (text, pp. 284–285)

2. Where is most of the body's iron found? What is the function of iron in hemoglobin and myoglobin? (text, p. 286)

3. How prevalent is iron deficiency? What populations are at high risk for iron deficiency? Define iron deficiency and iron-deficiency anemia. What are symptoms of iron-deficiency anemia? How does iron deficiency affect behavior? What is pica? (text, pp. 286–288)

4. What population is prone to iron overload? What are the symptoms of iron overload? How does excess iron relate to heart disease? (text, p. 289)

5. What is iron overload? What populations are at greatest risk for iron overload? (text, pp. 289; video)

6. Define *heme* and *nonheme* iron. (text, pp. 289–290)

7. What foods contain heme iron? What foods contain nonheme iron? What are the factors that enhance nonheme iron absorption? What are the factors that inhibit iron absorption? (text, pp. 289–291)

8. What are the symptoms of zinc deficiency? What form of retardation is associated with zinc deficiency? What are the effects of zinc toxicity with regard to zinc supplements? What foods are the richest sources of zinc? (text, pp. 291–292)

9. What is the role of selenium in the body? What condition found in children in China is associated with selenium deficiency? What food sources provide the most selenium? (text, pp. 292–293)

10. What is the function of fluoride in the body? What is the best source of fluoride? (text, pp. 293–294; video)

11. What condition results from fluoride deficiency? Define *fluorosis*. (text, pp. 293–294)

12. How does chromium function in the body? What are the best food sources of chromium? What is the controversy surrounding chromium supplements? (text, pp. 294–295)

13. What is the role of chromium? What are good food sources of chromium? What is the controversy surrounding chromium supplements? (text, pp. 294–295; video)

14. What function does copper serve in the body? What mineral interferes with copper absorption? What are the best food sources of copper? (text, p. 295)

15. What is the most common cause of iron deficiency worldwide? What are the symptoms of iron deficiency in adults? What populations are at greatest risk for iron deficiency? What foods are high in iron? What enhances the absorption of non-heme iron? (video)

16. What are phytochemicals? What are good sources of phytochemicals? Explain the link between lycopene and prostate cancer. What are good sources of lycopene? (video)

RELATED ACTIVITY

This activity is not required unless your instructor assigns it. It is offered as a suggestion to help you learn more about the material presented in this lesson.

Complete the form on the following page and return it to your instructor according to the established deadline.

- ❑ Fill in the Table of Trace Minerals

Fill in the Table of Trace Minerals

In the space provided, list up to three of the following: the major functions, deficiency disease and/or symptoms, toxicity symptoms, and the major food sources for the trace minerals. If a characteristic does not apply, write NA in the space.

Name of Mineral	Function(s) of Mineral	Deficiency Disease and/or Symptoms	Toxicity Symptoms	Major Food Sources
Iron	1. 2. 3.	1. 2. 3.	1. 2. 3.	1. 2. 3.
Zinc	1. 2. 3.	1. 2. 3.	1. 2. 3.	1. 2. 3.
Iodine	1. 2.	1. 2.	1. 2.	1. 2.
Selenium	1. 2.	1. 2.	1. 2.	1. 2.
Copper	1. 2.	1. 2.	1. 2.	1. 2.
Manganese	1. 2.	1. 2.	1. 2.	1. 2.
Fluoride	1. 2.	1. 2.	1. 2.	1. 2.
Chromium	1. 2.	1. 2.	1. 2.	1. 2.

PRACTICE TEST

The following items will help you evaluate your understanding of this lesson. Use the answer key at the end of the lesson to check your answers or to locate material related in each question.

Multiple Choice

Select the one choice that best answers the question.

1. What iron-containing compound carries oxygen in the bloodstream?
   A. Lipoproteins
   B. Myoglobin
   C. Antioxidants
   D. Hemoglobin

2. Your best friend complains that she is tired all of the time and has a problem concentrating. In addition to this, she likes to chew ice. Your friend is likely to have __________.
   A. potassium deficiency
   B. zinc deficiency
   C. magnesium deficiency
   D. iron deficiency

3. Which of the following must be available for thyroxine to be synthesized?
   A. Iron
   B. Iodine
   C. Zinc
   D. Fluoride

4. Which of the following population groups is LEAST susceptible to iron deficiency anemia?
   A. Older infants
   B. Children two to ten years of age
   C. Women of childbearing age
   D. Men twenty to forty-five years of age

5. What is the name given to the ingestion of nonnutritive substances?
   A. Pica
   B. Goiter
   C. Tetany
   D. Fluorosis

6. Which of the following is known to enhance iron absorption?
   A. Tea
   B. Coffee
   C. Foods containing vitamin C
   D. Foods containing vitamin E

7. Zinc is known to play an important role in all of the following functions EXCEPT __________.
   A. wound healing
   B. production of vitamin A
   C. production of sperm
   D. oxidation of fatty acids

8. Zinc is highest in foods that also contain a high amount of __________.
   A. fat
   B. fiber
   C. protein
   D. carbohydrate

9. What mineral, if taken in excess, is the number one cause of fatal accidental poisonings of American children under six years of age?
   A. Iron
   B. Selenium
   C. Chloride
   D. Potassium

10. Chromium deficiency is characterized by __________.
    A. high blood pressure
    B. high blood sugar
    C. enlargement of the liver
    D. enlargement of the thyroid gland

11. The most common cause of iron deficiency worldwide is __________.
   A. a lack of knowledge regarding food sources of iron
   B. not taking in enough vitamin B
   C. protein-energy malnutrition
   D. all of the above

12. One of the major hindrances of nonheme iron absorption is a naturally occurring substance called __________.
   A. iron binders
   B. the meat factor
   C. phytate
   D. MFP

13. One of the chief functions of chromium is participation in the metabolism of __________.
   A. iron
   B. proteins
   C. carbohydrates
   D. metallothionein

14. Fluoride deficiency is best known to lead to __________.
   A. dental decay
   B. osteoporosis
   C. discoloration of teeth
   D. nutritional muscular dystrophy

15. The most common cause of iron overload is __________.
   A. an injury to the GI tract
   B. a genetic predisposition
   C. excessive use of iron cookware
   D. excessive use of iron supplements

16. What term designates foods that contain nonnutrient substances which may provide health benefits beyond basic nutrition?
   A. Health foods
   B. Organic foods
   C. Phytochemicals
   D. Disease preventative foods

17. A good source of lycopene is __________.
    A. tomatoes
    B. hamburger meat
    C. celery
    D. carrots

Matching

Select the letter next to the word or phrase that best matches the numbered statements. Use answers only once.

______	18. Form of iron found only in animal flesh	A. Tea
______	19. Iodine deficiency or toxicity disease	B. Heme
______	20. Usual dietary source of fluoride	C. Goiter
		D. Anemia
		E. Nonheme

Essay Questions

Reflect on the items listed below; one or more may be included on the exam.

21. Explain the difference between *heme* and *nonheme* iron. How can the efficiency of absorption be increased for both types of iron?

22. Differentiate between *iron deficiency* and *iron-deficiency anemia* and describe the causes and symptoms of the latter.

ANSWER KEY

The following provides the answers and references for the practice test questions.

	Answer	Learning Objectives	References
1.	D	LO 1	Sizer/Whitney, p. 286
2.	D	LO 2	Sizer/Whitney, pp. 286–287
3.	B	LO 6	Sizer/Whitney, p. 284–285
4.	D	LO 2	Sizer/Whitney, p. 288
5.	A	LO 2	Sizer/Whitney, p. 287
6.	C	LO 4	Sizer/Whitney, pp. 289–290
7.	D	LO 6	Sizer/Whitney, p. 291
8.	C	LO 9	Sizer/Whitney, p. 292
9.	A	LO 3	Sizer/Whitney, p. 289
10.	B	LO 7	Sizer/Whitney, pp. 294–295
11.	A	LO 2	Video
12.	C	LO 4	Video
13.	C	LO 6	Video
14.	A	LO 9	Video
15.	B	LO 3	Video
16.	C	LO 10	Video
17.	A	LO 10	Video
18.	B	LO 4	Sizer/Whitney, p. 289
19.	C	LO 7, 8	Sizer/Whitney, p. 285
20.	A	LO 9	Sizer/Whitney, p. 294
21.		LO 4	Sizer/Whitney, pp. 289–291
22.		LO 2	Sizer/Whitney, pp. 287–288

Lesson 16

Physical Activity: Fitness Basics

OVERVIEW

Although the primary focus of this telecourse is the acquisition of nutrition knowledge leading toward a healthier life, the equation is only partially complete. The inclusion of physical activity as a regular part of one's life provides one more component of the equation leading toward total well-being. This lesson will show how physical activity can have a significant, positive influence on your overall health, whether you are a seasoned veteran of activity, a novice, or someone who has never been very active. Combined with appropriate nutrition choices and other positive lifestyle changes, physical activity is nearly the "magic bullet" we all seem to be seeking for ultimate health and well-being. You will learn how to incorporate your personal goals into a program that emphasizes more activity in your life. You will debunk the myths associated with physical activity by learning what fuels are best suited for specific activities.

LEARNING OBJECTIVES

Upon completing this lesson, you should be able to:

1. Cite the benefits associated with being physically fit.
2. List the American College of Sports Medicine (ACSM) guidelines for developing and maintaining physical fitness.
3. Define the components of a sound fitness program.
4. Explain how glucose is used during aerobic and anaerobic exercise.
5. Explain how fat is used during exercise.
6. Summarize how the body adjusts its fuel mix to respond to physical activity of varying intensity levels and duration.
7. Explain how activity and good nutrition can be incorporated into one's life, whether at work or at home.

LESSON ASSIGNMENTS

Review the following assignment in order to schedule your time appropriately. Pay careful attention. The titles and numbers of the textbook chapter, the telecourse guide lesson, and the video program may be different from one another.

Text: Sizer and Whitney, *Nutrition: Concepts and Controversies*, Chapter 10, "Nutrients, Physical Activity, and the Body's Responses," pp. 359–392

Video: "Physical Activity: Fitness Basics" from the series *Nutrition Pathways*

LESSON FOCUS POINTS

The following questions are designed to help you get the most out of the lesson. After you read the textbook assignments and view the video, you should be able to answer these questions.

1. Explain how being more physically fit impacts the following: sleep, nutritional health, body composition, bone density, immunity, risk for cancer, circulation/lung function, risk for cardiovascular disease, risk for diabetes, anxiety/depression, self-image, and longevity. (text, pp. 360–361)

2. What are the American College of Sports Medicine (ACSM) guidelines for developing and maintaining physical fitness? (text, p. 361)

3. Define the following components of a sound fitness program: flexibility, muscle strength, muscle endurance, and cardiorespiratory endurance. Explain the following principles of conditioning: progressive overload, frequency, intensity, duration, warm-up, and cool-down. (text, p. 362)

4. Define *aerobic* and *anaerobic*. What is lactic acid and when is it produced? What is the primary fuel used during extremely intense exercise of short duration? How soon will glycogen depletion occur after the onset of intense activity? What do marathon runners refer to the point of total glucose depletion? What are four ways to maintain adequate blood glucose during endurance exercise? What is "carbohydrate loading"? (text, pp. 362; 366–369)

5. What are the benefits of weight training? What are the benefits of aerobic training for cardiorespiratory endurance? For muscle conditioning? (text, pp. 362–366)

6. What are three preliminary steps to take prior to beginning an exercise program? Why would a person with high blood cholesterol consider participating in a regular, aerobic exercise program? (text, p. 363; video)

7. What factors affect the type of fuel used by muscles during activity? What is the primary fuel used during the twenty minutes of moderate exercise? What fuel is increasingly used by muscles after twenty minutes of activity? Does an athlete continue to burn more energy *at rest* than an inactive person or casual exerciser? (text, pp. 366–371)

8. What are the factors that affect fat use during physical activity? What happens to fat utilization as exercise intensity increases? How does training affect fat use? (text, pp. 370–371)

9. Why is walking considered one of the best types of exercise to perform? What exercise factors positively impact health and longevity the most? (video)

10. What populations are at increased risk for dying from all causes? (video)

11. What does the latest research state regarding ultra-athletes or marathoners and immune function? (video)

12. What advantages does an on-site fitness program provide to employees? How does exercise affect the brain? (video)

13. Describe a general but adequate diet to support activity. (video)

KEY CONCEPTS

Physical Activity Guidelines

Guidelines for developing and maintaining physical fitness:

- Frequency of activity: three to five days per week
- Intensity of activity: 55 to 90 percent maximum heart rate
- Duration of activity: twenty to sixty minutes of continuous activity
- Mode of activity: any activity that uses large muscle groups
- Resistance activity: strength training of moderate intensity at least two times per week
- Flexibility activity: stretching of the major muscle groups two to three times a week

Benefits of Physical Fitness

- Reduced risk of cardiovascular diseases
- Reduced risk of some types of cancer (especially of the colon or breast)
- Improved mental outlook and lessened likelihood of depression
- Improved mental functioning
- Feeling of vigor
- Feeling of belonging—the fun and companionship of sports
- Strong self-image and self-confidence
- Reduced body fatness and increased lean tissue
- Greater bone density and, thus, lessened risk or reduced severity of bone disease later in life
- Sound, beneficial sleep
- A more youthful appearance, healthy skin, and improved muscle tone
- Faster wound healing
- Improvement or elimination of menstrual cramping
- Improved resistance to colds and infection

Two Types of Exercise

I. Aerobic exercise (with oxygen)

- Lower intensity
- Endurance-oriented
- Fat-burning

Examples:

- Long-distance running
- Walking
- Aerobics

II. Anaerobic exercise (without oxygen)

- Higher intensity
- Strength/speed-oriented
- Carbohydrate-burning

Examples:

- Sprinting
- Football
- Weight-lifting

Glucose and Fatty Acids in Their Energy-Releasing Pathways

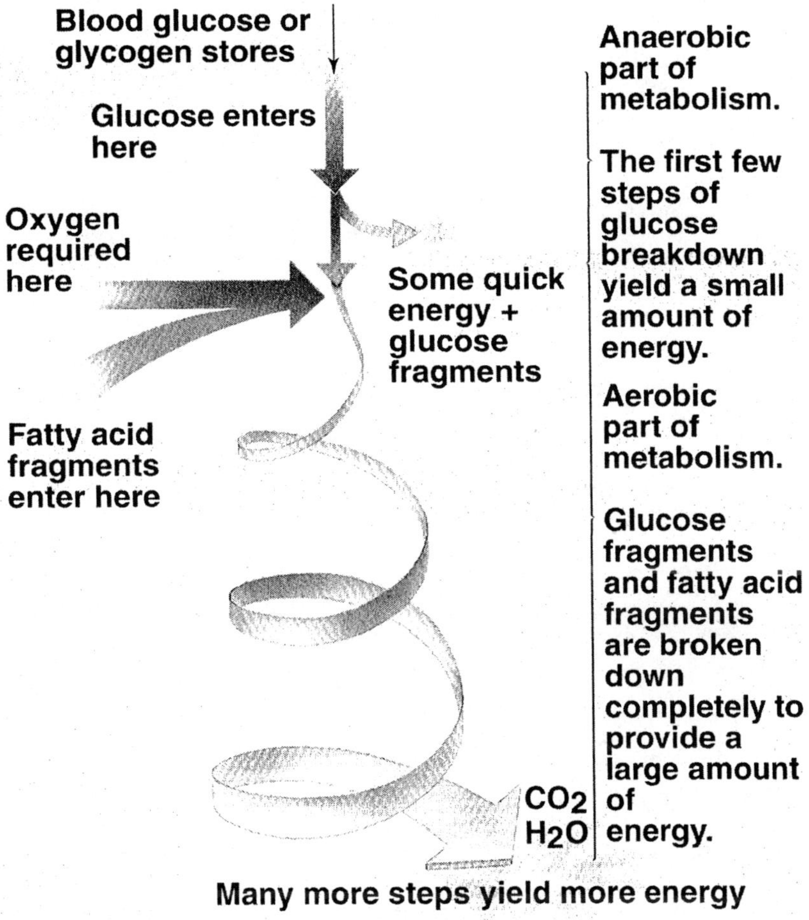

© 2000 Wadsworth Publishing Company/ITP

PRACTICE TEST

The following items will help you evaluate your understanding of this lesson. Use the answer key at the end of the lesson to check your answers or to locate material related in each question.

Multiple Choice

Select the one choice that best answers the question.

1. According to the American College of Sports Medicine, which of the following would meet the exercise schedule to maintain an appropriate level of fitness?
   A. Two hours of aerobic exercise daily
   B. One hour of strength training four times a week
   C. Five minutes of aerobic exercise two times a week
   D. A minimum of thirty minutes of any continuous activity using large muscle groups on most days of the week

2. Muscles respond to the overload of exercise by gaining __________.
   A. strength
   B. fat
   C. size
   D. A and C

3. The effect of regular exercise on heart and lung function is known as __________.
   A. muscle fitness
   B. muscle endurance
   C. cardiopulmonary adaptation
   D. cardiorespiratory conditioning

4. Anaerobic glucose breakdown produces fragments of glucose molecules that accumulate in the tissues and blood called __________.
   A. urea
   B. lactic acid
   C. uric acid
   D. amino acid

5. What is the predominant fuel used by muscle cells during low or moderate intensity activity?
   A. Fat
   B. Protein
   C. Glycogen
   D. Blood glucose

6. How much time is usually needed in vigorous activity to cause depletion of glycogen reserves?
   A. One hour
   B. Two hours
   C. Three hours
   D. Four hours

7. A person who continues exercising moderately for longer than _____ minutes begins to use less glucose and more fat for fuel.
   A. five
   B. ten
   C. fifteen
   D. twenty

8. When a marathon runner experiences the phenomenon known as "hitting the wall," what nutrient is most likely depleted?
   A. Water
   B. Protein
   C. Glucose
   D. Fatty acids

9. Research clearly shows that the risk of dying not only from heart disease but from all causes is greatest for __________.
   A. elite athletes such as marathoners
   B. sedentary people, also called "couch potatoes"
   C. occasional exercisers, also called "weekend warriors"
   D. moderately active people who exercise three to four days per week

10. Individuals who run 60 or 70 miles per week, such as marathon runners, may actually __________.
   A. feel it is not enough to achieve their goals
   B. be overexercising to the benefit of their health
   C. suppress their immune function and cause disease
   D. all of the above

11. In addition to being able to exercise at work, an employee who has access to a wellness program and fitness facility at the job site __________.
   A. is more likely to make positive changes in lifestyle
   B. has less absenteeism from work
   C. becomes more educated in health issues
   D. all of the above

12. Regardless of whether the intensity of an exercise program is low, moderate, or vigorous, the best diet emphasizes __________.
   A. protein
   B. fat
   C. carbohydrate
   D. water

13. Walking is considered the best form of exercise because it __________.
   A. is not expensive
   B. promotes cardiovascular endurance
   C. is easy to perform
   D. all of the above

14. A person with high blood cholesterol should consider participating in an aerobic exercise program such as walking or jogging because aerobic exercise __________.
   A. will raise HDL ("good") cholesterol after a period of training and improve the risk for heart problems
   B. will cause a sudden increase in the heartrate to near maximal levels which lowers the LDL ("bad") cholesterol
   C. forces more LDL ("bad") cholesterol out of the body through waste elimination
   D. keeps the blood circulating faster, even at rest, therefore causing a decrease in total blood cholesterol

Matching

Select the letter next to the word or phrase that best matches the numbered statements. Use answers only once.

_____	15. The capacity of the joints to move with less chance of injury	A. Strength
		B. Oxygen
_____	16. The capacity of the heart and lungs to sustain vigorous activity	C. Flexibility
_____	17. Required for aerobic metabolism	D. Cardiorespiratory endurance
_____	18. The ability of muscles to exert maximal forces	E. Carbon dioxide

Essay Questions

Reflect on the items listed below; one or more may be included on the exam.

19. Discuss the physiological and psychological benefits of being physically fit.

20. Identify the characteristics and benefits of cardiorespiratory endurance.

ANSWER KEY

The following provides the answers and references for the practice test questions.

	Answer	Learning Objectives	References
1.	D	LO 2	Sizer/Whitney, pp. 361–362
2.	D	LO 3	Sizer/Whitney, p. 362
3.	D	LO 3	Sizer/Whitney, p. 362
4.	B	LO 4	Sizer/Whitney, p. 367
5.	A	LO 5, 6	Sizer/Whitney, p. 370
6.	B	LO 4	Sizer/Whitney, p. 368
7.	D	LO 6	Sizer/Whitney, p. 368
8.	C	LO 4	Sizer/Whitney, pp. 368–369
9.	B	LO 1	Video
10.	C	LO 2	Video
11.	D	LO 7	Video
12.	C	LO 7	Video
13.	D	LO 7	Video
14.	A	LO 1	Video
15.	C	LO 3	Sizer/Whitney, p. 362
16.	D	LO 3	Sizer/Whitney, p. 362
17.	B	LO 4	Sizer/Whitney, p. 362
18.	A	LO 3	Sizer/Whitney, p. 362
19.		LO 1	Sizer/Whitney, pp. 360–361
20.		LO 3	Sizer/Whitney, pp. 362–364

Lesson 17

Physical Activity: Beyond Fitness

OVERVIEW

Lesson 17 will go beyond the basic information you need to begin being more active. In this lesson you will learn how the body uses food to fuel activities, how specific nutrients support activity, how water and electrolytes affect the athlete, what is an acceptable diet for a physically active person, and how sports drinks and other ergogenic aids benefit or hinder athletic performance.

LEARNING OBJECTIVES

Upon completing this lesson, you should be able to:

1. Compare and contrast how protein is used during both physical activity and rest.
2. Cite the recommended protein intake for athletes.
3. Describe the use of vitamins, minerals, and supplements during activity.
4. State how fluids and electrolytes affect endurance activity.
5. Cite the recommended fluid intake for physical activity.
6. Describe a diet to support fitness including pre- and post-competition meals.
7. List the risks of using ergogenic aids to improve athletic performance.
8. Explain the importance of diet and exercise for physically challenged people and people with high blood lipids.

LESSON ASSIGNMENT

Review the following assignment in order to schedule your time appropriately. Pay careful attention. The titles and numbers of the textbook chapter, the telecourse guide lesson, and the video program may be different from one another.

Text: Sizer and Whitney, *Nutrition: Concepts and Controversies*, Chapter 10, "Nutrients, Physical Activity, and the Body's Responses," pp. 359–392

Video: "Physical Activity: Beyond Fitness" from the series *Nutrition Pathways*

LESSON FOCUS POINTS

The following questions are designed to help you get the most out of the lesson. After you read the textbook assignments and view the video, you should be able to answer these questions.

1. What happens to body protein synthesis during activity? During active muscle-building training, how much muscle may an athlete add each day to body protein? How much protein is typically used as a fuel source? Explain how the following factors affect protein use during exercise: diet, intensity, duration, and training. (text, pp. 371–372)

2. How much protein is recommended for active people? How much protein is recommended for power athletes? How much protein is recommended for endurance athletes? (text, pp. 372–373)

3. Why do athletes require more antioxidant vitamins? In general, does an athlete need to take a vitamin-mineral supplement? What mineral deficiency is a potential risk for female athletes? What is "sports anemia"? (text, pp. 373–375)

4. What is the first symptom of dehydration in an athlete? What is the best fluid for an everyday, active person? For endurance athletes? Why should athletes be concerned about caffeine and alcohol intake? What are three advantages athletes receive from consuming sports drinks? (text, pp. 376–379)

5. How much water should you drink to replace one pound of weight lost during exercise? What is the recommended hydration schedule during activity? (text, pp. 376–379)

6. What is the basis of a healthy diet for an active person? How many calories are recommended for a pregame meal? How many hours before competition should the meal be eaten? What foods are recommended for a pregame meal? What foods are recommended after competition? (text, pp. 380–383)

7. What substance is depleted during repetitive exercises? What foods should be the foundation of a diet for an athlete? How much food should be eaten the day of a competition? Why are high carbohydrate foods recommended the day before a competition? What advice is given regarding caffeine intake prior to exercise? (text, pp. 380–383; video)

8. Define ergogenic aid. What is DHEA (dehydroepiandrosterone)? What are its side effects? (text, pp. 385–391)

9. Explain how exercise and nutrition impact a physically challenged person. (video)

10. What are ergogenic aids? How do nutrition supplements impact the performance of athletes? What are some of the side effects of steroid use among athletes? What is epoetin (EPO)? What advice is given regarding the use of sports drinks and performance? (video)

11. How can regular exercise impact energy, stress, and blood lipid levels? (video)

KEY CONCEPTS

Hydration Schedule for Physical Activity

When to Drink	Approximate Amount of Fluid
Two hours before exercise	3 cups
Ten to fifteen minutes before exercise	2 cups
Every fifteen minutes during exercise	1 cup
After exercise	*2 cups

*Per pound of body weight lost

©1999 Wadsworth Publishing Company/ITP

PRACTICE TEST

The following items will help you evaluate your understanding of this lesson. Use the answer key at the end of the lesson to check your answers or to locate material related in each question.

Multiple Choice

Select the one choice that best answers the question.

1. Protein contributes an average of about _____ percent of the total fuel used during activity and during rest.
   A. 5
   B. 10
   C. 15
   D. 20

2. The American Dietetic Association recommends ________ grams(s) of protein per kilogram of body weight each day for an athlete.
   A. 0.8 to 1.0
   B. 1.0 to 1.5
   C. 1.5 to 2.0
   D. 2.0 to 2.5

3. The most important antioxidant vitamin related to physical activity is __________.
   A. vitamin E
   B. beta carotene
   C. vitamin C
   D. folate

4. Caffeine use in athletes __________.
   A. has a potentially dangerous diuretic effect
   B. slows down reactions
   C. depletes glycogen stores
   D. has no adverse effects

5. Sports anemia __________.
   A. is a normal adaptation to physical activity
   B. requires treatment with iron supplements
   C. is temporary
   D. A and B
   E. A and C

6. The first sign of dehydration is typically __________.
   A. fatigue
   B. dizziness
   C. intense thirst
   D. intense sweating

7. The digestive system can absorb about one _________ of fluid an hour.
   A. cup
   B. pint
   C. quart
   D. gallon

8. Which of the following would be the best choice for the typical athlete who needs to rehydrate?
   A. Liquid meal replacer
   B. Salt tablets and tap water
   C. Diluted juice or cool water
   D. Water warmed to body temperature

9. What is the recommended composition of the pregame meal of the athlete?
   A. Low protein
   B. High protein
   C. Low carbohydrate
   D. High carbohydrate

10. Whether performing an anaerobic sport, such as diving, or an aerobic sport, such as marathon running, the training routine __________.
    A. involves repetitive use of muscles during the exercise
    B. depletes muscle glycogen
    C. must be supported by a high carbohydrate diet
    D. all of the above

11. Regardless of the physical pursuits of an athlete, the recommended diet should be a foundation of __________.
    A. protein
    B. amino acids
    C. water
    D. complex carbohydrates

12. Physical activity is very important to physically challenged people because __________.
    A. it keeps them from becoming bored
    B. they can eat more and, by consuming more nutrients, become healthier
    C. they are not as strong as people who are not physically challenged
    D. all of the above

13. An illegal drug used by athletes as an alternative for blood doping is _______.
    A. epoetin
    B. an anabolic steroid
    C. effective, as long as it is taken in moderation
    D. all of the above

14. Which of the following can be improved through a regular aerobic exercise program?
    A. Triglycerides and HDL levels
    B. Total blood cholesterol
    C. LDL levels
    D. All of the above

Fill in the Blank

Insert the correct word or words in the blank for each item.

15. The foundation of an athlete's diet should consist of _________________.

16. The ____________ component of a blood lipid profile has been shown to be improved by regular aerobic exercises.

Essay Questions

Reflect on the items listed below; one or more may be included on the exam.

17. Describe the composition and characteristics of an ideal pregame meal.

18. Discuss the need for water in maintaining physical performance. What are the symptoms of dehydration? What are the recommendations for ensuring that the body is well-hydrated prior to an athletic event?

ANSWER KEY

The following provides the answers and references for the practice test questions.

	Answer	Learning Objectives	References
1.	B	LO 1	Sizer/Whitney, p. 372
2.	B	LO 2	Sizer/Whitney, p. 373
3.	A	LO 3	Sizer/Whitney, p. 374
4.	A	LO 7	Sizer/Whitney, p. 379
5.	E	LO 3	Sizer/Whitney, p. 375
6.	A	LO 4	Sizer/Whitney, p. 376
7.	C	LO 4, 5	Sizer/Whitney, p. 376
8.	C	LO 4	Sizer/Whitney, p. 377
9.	D	LO 6	Sizer/Whitney, p. 382
10.	D	LO 6	Video
11.	D	LO 6	Video
12.	B	LO 8	Video
13.	A	LO 7	Video
14.	A	LO 8	Video
15.	complex carbohydrates	LO 6	Video
16.	HDL (or triglycerides)	LO 8	Video
17.		LO 6	Sizer/Whitney, pp. 382–383
18.		LO 4	Sizer/Whitney, pp. 376–379

Lesson 18

Life Cycle: Pregnancy

OVERVIEW

As you enter different stages of growth and development throughout the life cycle, your nutritional needs will vary. You will require the same nutrients as someone much older or much younger than yourself, but the amount needed will be unique to the stage of life. This lesson will focus on nutrition needs as a woman prepares for pregnancy, as well as how nutrition affects fetal growth and development. You will see how exercise plays a part as maternal needs change during pregnancy, how weight gain impacts pregnancy, and how teen pregnancy impacts the mother as well as the developing fetus.

LEARNING OBJECTIVES

Upon completing this lesson, you should be able to:

1. Describe the function of the placenta.
2. Discuss the critical periods of fetal development.
3. Discuss the increase in energy and protein needs during pregnancy.
4. Discuss the increased need for folate and B_{12} during pregnancy.
5. Discuss the increased need for calcium and iron during pregnancy.
6. Explain how maternal weight affects pregnancy.
7. Cite the current recommendations for weight gain during pregnancy.
8. Describe appropriate exercise guidelines during pregnancy.
9. Discuss the risks associated with adolescent pregnancy.

10. Describe nutritional strategies for relieving common pregnancy-related discomforts.

11. Specify how various drugs impact pregnancy.

LESSON ASSIGNMENT

Review the following assignment in order to schedule your time appropriately. Pay careful attention; the titles and numbers of the textbook chapter, the telecourse guide lesson, and the video program may be different from one another.

Text: Sizer and Whitney, *Nutrition: Concepts and Controversies*, Chapter 13, "Life Cycle Nutrition: Mother and Infant," pp. 481–518

Video: "Life Cycle: Pregnancy" from the series *Nutrition Pathways*

LESSON FOCUS POINTS

The following questions are designed to help you get the most out of the lesson. After you read the textbook assignments and view the video, you should be able to answer these questions.

1. What is the effect of malnutrition on fetal development? Define low birthweight. What complications are low-birthweight infants likely to experience? What percentage of infant deaths before one year of age are attributed to low birthweight? What is the function of the placenta? What attaches the placenta to the fetus? (text, pp. 482–484)

2. What developments occur in the embryo at eight weeks after conception? What is a fetus? What are "critical periods" during pregnancy? (text, pp. 482–484)

3. How many extra calories per day does a pregnant woman need? What is the RDA for protein during pregnancy? What roles do folate and B_{12} play in a successful pregnancy? What role does calcium play during the early stages of pregnancy? What role does iron play? What nutrient is typically supplemented during the second and third trimesters for all women? At what point during pregnancy do neural tube defects occur? What is spina bifida? What nutrient

can prevent neural tube defects? When should a woman use folate supplementation? (text, pp. 485–490)

4. To what is the primary weight gain during pregnancy attributed? What advice is given regarding dieting during pregnancy? (text, p. 491; video)

5. What is the greatest risk underweight women have during pregnancy? What medical complications can occur in an overweight woman during pregnancy? Cite the recommended weight gain for the following women during pregnancy: healthy weight, underweight, and overweight women. What contributes to weight gain during pregnancy? (text, pp. 491–492)

6. What advice is given to pregnant women who want to begin an exercise program? Why is swimming considered the best exercise for pregnant women? (text, p. 492; video)

7. What are the benefits of exercise during pregnancy? Cite the guidelines for exercise during pregnancy. (text, pp. 492–493)

8. Cite the risks associated with teenage pregnancy. How much weight gain is recommended for young teenage mothers? (text, p. 493)

9. What seems to be responsible for early pregnancy nausea? What nutritional strategies will alleviate nausea? What nutritional strategies can prevent or alleviate constipation? How can heartburn be relieved? What is pica? (text, p. 493)

10. What is the most at-risk age group for pregnancy? What is the greatest risk for teenage pregnancies? What is the recommendation for iron supplementation for pregnant women? What is the recommendation for folic acid supplementation for pregnant women? What condition will the intake of folic acid prevent? (text, p. 493; video)

11. What are the risks associated with drug use and fetal development? What risks are associated with smoking? What is sudden infant death syndrome? What is the effect of dieting on fetal development? How does alcohol intake affect pregnancy? What is fetal alcohol syndrome? What is the recommendation regarding caffeine intake during pregnancy? (text, pp. 495–496)

12. What is the recommendation regarding caffeine intake during pregnancy? (text, p. 496)

13. What is gestational diabetes? What is pregnancy-induced hypertension? What is preeclampsia? How does edema of preeclampsia differ from normal edema of pregnancy? What is eclampsia? (text, pp. 499–500)

14. Describe how being a single, working mother can impact personal fitness and nutrition goals. (video)

KEY CONCEPTS

Effects of Nutrient Deficiencies During Pregnancy

Nutrient	Deficiency Effect
Energy	Low infant birthweight
Protein	Reduced infant head circumference
Folate	Miscarriage and neural tube defect
Vitamin D	Low infant birthweight
Calcium	Decreased infant bone density
Iron	Low infant birthweight and premature birth
Iodine	Cretinism (varying degrees of mental and physical retardation in the infant)
Zinc	Congenital malformations

Source: Adapted from L. K. DeBruyne and S. R. Rolfes, Life Cycle Nutrition: Conception Through Adolescence, ed. E. N. Whitney (St. Paul, MN: West, 1989), p. 68.

PRACTICE TEST

The following items will help you evaluate your understanding of this lesson. Use the answer key at the end of the lesson to check your answers or to locate material related in each question.

Multiple Choice

Select the one choice that best answers the question.

1. What organ of the pregnant woman is central to the exchange of nutrients for waste products with the fetus?
   A. Uterus
   B. Vagina
   C. Placenta
   D. Amniotic sac

2. Which of the following nutrients is related to the prevention of neural tube defects?
   A. Vitamin B_6
   B. Vitamin B_{12}
   C. Folate
   D. Zinc

3. What is the recommended range of weight gain during pregnancy for a normal-weight woman?
   A. 10 to 18 pounds
   B. 19 to 24 pounds
   C. 25 to 35 pounds
   D. 35 to 40 pounds

4. All pregnant women need generous amounts of __________ to spare their protein and to provide energy.
   A. carbohydrates
   B. fat
   C. protein
   D. calories

5. What is the recommended *increase* in energy intake for the second and third trimester of pregnancy?
   A. 200 calories per day
   B. 300 calories per day
   C. 340 calories per day
   D. 440 calories per day

6. A pregnant teenager with a body mass index in the normal range is encouraged to gain _____ pounds or so.
   A. 13
   B. 20
   C. 35
   D. 42

7. Which of the following is one of the recommendations to treat pregnancy-associated heartburn?
   A. Eat many small meals
   B. Drink fluids only with meals
   C. Exercise within thirty minutes after eating
   D. Lie down within fifteen minutes after eating

8. The weight that pregnant women put on is nearly all lean tissue.
   A. True
   B. False

9. All of the following statements are true concerning the placenta EXCEPT __________.
   A. it is a sort of cushion in which the mother's and baby's blood vessels intertwine
   B. it allows for the exchange of materials between baby and mother
   C. the fetus receives nutrients and carbon dioxide across the placenta
   D. the mother's blood picks up the baby's waste material through the placenta and excretes them

10. Physical activity during pregnancy can prevent complications in pregnancy, facilitate labor, and reduce psychological stress.
    A. True
    B. False

11. A pregnant woman needs how many grams of protein per day?
    A. 30
    B. 45
    C. 50
    D. 60

12. Early malnutrition in the mother affects the baby's __________.
    A. heart
    B. brain
    C. lungs
    D. A and B

13. A sudden and large weight gain during pregnancy is always a danger signal.
    A. True
    B. False

14. Which of the following should be avoided during pregnancy?
    A. alcohol
    B. smoking
    C. caffeine consumption
    D. A and B

15. Advice to pregnant women who want to exercise includes __________.
    A. do not begin a new program
    B. swimming, but not weight-lifting
    C. walking, but not too intensely
    D. all of the above

16. Dieting during pregnancy is __________.
    A. recommended if the mother is overweight
    B. not recommended at all
    C. dependent upon the initial weight of the mother
    D. directly correlated to high birthweight

17. The recommendation for iron supplementation is __________.
    A. greater for pregnant teens
    B. about 30 to 60 milligrams per day
    C. necessary to preserve iron stores
    D. all of the above

18. One of the reasons young mothers lose sight of personal fitness or nutrition goals is due to __________.
    A. overwork and fatigue
    B. the stresses of being a new mother
    C. postpartum depression
    D. all of the above

Essay Questions

Reflect on the items listed below; one or more may be included on the exam.

19. What are several benefits of exercise specifically for the pregnant woman? What types of exercise should be avoided and why?

20. Describe the risks associated with adolescent pregnancy.

ANSWER KEY

The following provides the answers and references for the practice test questions.

	Answer	Learning Objectives	References
1.	C	LO 1	Sizer/Whitney, p. 482
2.	C	LO 4	Sizer/Whitney, p. 488
3.	C	LO 7	Sizer/Whitney, p. 491
4.	A	LO 3	Sizer/Whitney, pp. 485–487
5.	C	LO 3	Sizer/Whitney, p. 485
6.	C	LO 9	Sizer/Whitney, p. 493
7.	A	LO 10	Sizer/Whitney, p. 494
8.	A	LO 6	Sizer/Whitney, p. 491
9.	C	LO 2	Sizer/Whitney, p. 482
10.	A	LO 8	Sizer/Whitney, p. 492
11.	D	LO 3	Sizer/Whitney, p. 487
12.	D	LO 2	Sizer/Whitney, p. 484
13.	A	LO 6	Sizer/Whitney, p. 491
14.	D	LO 11	Sizer/Whitney, pp. 494–497
15.	D	LO 8	Video
16.	B	LO 6, 7	Video
17.	D	LO 5	Video
18.	D	LO 8	Video
19.		LO 8	Sizer/Whitney, p. 492
20.		LO 9	Sizer/Whitney, p. 493

Lesson 19

Life Cycle: Lactation and Infancy

OVERVIEW

In this lesson you will investigate the advantages and disadvantages of both breastfeeding and infant formula and why women select the method of feeding that is best for them and their infants. You will discover circumstances which may prevent a mother from breastfeeding; learn why and when cow's milk should be introduced to the infant; and you will learn when to introduce solid foods and what foods are best to introduce to the infant.

LEARNING OBJECTIVES

Upon completing this lesson, you should be able to:

1. List the benefits of breastfeeding to both the infant and to the mother.
2. Discuss the nutrient requirements of the lactating woman.
3. Explain the effects of stress on lactation.
4. Identify substances and conditions which may prohibit breastfeeding.
5. Describe the energy and nutrient needs of the infant.
6. Compare and contrast breast milk with infant formula and with cow's milk.
7. Describe the process of introducing solid foods to the infant.

LESSON ASSIGNMENT

Review the following assignment in order to schedule your time appropriately. Pay careful attention; the titles and numbers of the textbook chapter, the telecourse guide lesson, and the video program may be different from one another.

Text: Sizer and Whitney, *Nutrition: Concepts and Controversies*, Chapter 13, "Life Cycle Nutrition: Mother and Infant," pp. 481–518

Video: "Life Cycle: Lactation and Infancy" from the series *Nutrition Pathways*

LESSON FOCUS POINTS

The following questions are designed to help you get the most out of the lesson. After you read the textbook assignments and view the video, you should be able to answer these questions.

1. In order to produce milk, how many extra calories per day must a lactating woman take in during the first six months of lactation? What is the effect of nutrient deficiencies on the quantity and quality of breast milk? What is the recommendation for fluids during lactation? Why might iron supplements be necessary for lactating women? What foods might affect the flavor of breast milk? (text, pp. 500–501)

2. How does stress affect lactation or breast milk production? (text, pp. 500–501)

3. Describe the effects of the following on breast milk production and/or infant development: alcohol, caffeine, smoking, and viruses. (text, pp. 501–502)

4. How many calories per day do infants require based on body weight? What is the most important nutrient for infants? (text, p. 503)

5. What protection is provided to an infant who drinks mother's milk? What physiological and other benefits are provided to a woman who breastfeeds? (text, pp. 505–506)

6. Why is adequate iron and Vitamin C intake important for an infant? What is the carbohydrate found in breast milk and what is its function? What is the chief protein found in breast milk? What vitamin is typically deficient in breast milk? What supplements are necessary for breastfed infants? Compare the nutrients of breast milk with infant formula for iron and protein content. Why is cow's milk inappropriate during the first year? (text, pp. 505–508)

7. Compare the energy needs of a breastfed infant to a formula-fed infant. (text, pp. 505–508; video)

8. At what age should solid foods be introduced to the infant? What are the indications that solid foods may be given to an infant? Why should foods be introduced one at a time? What solid food is the first one generally introduced to an infant and why? (text, pp. 509–511)

9. What is the recommended age to introduce cow's milk? At what age should breastfed infants begin to receive iron supplementation? (text, pp. 509–510; video)

10. What are the guidelines that indicate a breastfed infant is getting enough food and calories? (video)

11. What are the nutritional requirements for a woman who is not breastfeeding? (video)

12. What is the recommendation regarding introducing fruit versus vegetables? (video)

13. What is the "letdown" reflex? How does stress affect the letdown reflex? (video)

PRACTICE TEST

The following items will help you evaluate your understanding of this lesson. Use the answer key at the end of the lesson to check your answers or to locate material related in each question.

Multiple Choice

Select the one choice that best answers the question.

1. Characteristics of colostrum include the following:
   A. It contains antibodies from the mother's blood.
   B. It contains white cells from the mother's blood.
   C. It may transmit a bacterial infection to the infant if the mother has one.
   D. A and B

2. The effect of maternal nutritional deprivation during lactation is to _________.
   A. reduce the quality of milk
   B. reduce the quantity of milk
   C. make breastfeeding impossible
   D. A and B

3. Under which of the following circumstances would it still be acceptable for a mother to breastfeed?
   A. She has alcohol abuse.
   B. She has a drug addiction.
   C. She has an ordinary cold.
   D. She has a communicable disease.

4. Characteristics of breast milk include __________.
   A. its sodium content is low
   B. its protein is largely alpha-lactalbumin
   C. its vitamin C content is poor
   D. A and B

5. A nursing mother produces about _____ ounces of milk a day
   A. 10
   B. 15
   C. 20
   D. 25

6. Which of the following should NOT be used to feed an infant?
   A. Whole milk
   B. Ready-to-feed formula
   C. Liquid concentrate formula appropriately diluted
   D. Powdered formula or evaporated milk formula appropriately reconstituted.

7. It is desirable to begin feeding the infant iron-fortified cereals by about _____ months.
   A. three
   B. six
   C. nine
   D. twelve

8. Whole milk should not be given routinely to a child until after the age of __________.
   A. two weeks
   B. three months
   C. one year
   D. six years

9. Compared to formula-fed infants, breastfed infants seem to need __________.
   A. less energy
   B. more energy
   C. more iron
   D. less iron

10. One of the ways you can tell if an infant is getting enough food is __________.
    A. by the number of wet diapers per day
    B. if the infant is gaining two pounds per week
    C. if the infant sleeps through the night at three weeks
    D. by the number of feedings per day

11. A woman who is not breastfeeding __________.
   A. has the same nutrient requirements as nonlactating women
   B. needs more nutrients than nonlactating women
   C. has the same nutrient requirements as lactating women
   D. has a greater need for water than lactating women

12. Nutrition scientists recommend that cow's milk be given to infants __________.
   A. as soon as they can sit up
   B. after the age of twelve months
   C. who are eating solid foods
   D. any of the above

13. Some research shows that introducing fruits before vegetables __________.
   A. will cause the infant to develop a sweet tooth
   B. causes more allergic reactions
   C. doesn't really matter; you can introduce either one first
   D. causes excess weight gain in the infant

14. The order of introduction of fruits and vegetables to infants __________.
   A. doesn't seem to matter
   B. should be to start with bananas
   C. should be to start with squash
   D. should be to mix them together at the same meal

15. Stress in the mother who breastfeeds has been shown to __________.
   A. inhibit the quality of milk produced
   B. cause greater milk production
   C. inhibit the letdown reflex
   D. cause weight loss in the infant

Fill in the Blank

Insert the correct word or words in the blank for each item.

16. At least one criteria that indicates whether an infant is getting enough food is

   ____.

Matching

Select the letter next to the word or phrase that best matches the numbered statements. Use answers only once.

_____	17. Typical daily energy need, in calories per kilogram body weight, of an infant	A. 100
		B. 120
_____	18. Process whereby breast milk is gradually replaced by formula or semisolid foods	C. Weaning
_____	19. Nutrient that is low in cow's milk but abundant in human milk	D. Vitamin B
		E. Vitamin C

Essay Questions

Reflect on the items listed below; one or more may be included on the exam.

20. Compare and contrast major differences in nutrient content of breast milk and cow's milk.

21. Why is the risk of dehydration higher for infants than for adults?

ANSWER KEY

The following provides the answers and references for the practice test questions.

	Answer	Learning Objectives	References
1.	D	LO 1	Sizer/Whitney, p. 506
2.	B	LO 1	Sizer/Whitney, p. 500
3.	C	LO 2	Sizer/Whitney, pp. 501–502
4.	D	LO 4	Sizer/Whitney, pp. 505–506
5.	D	LO 2	Sizer/Whitney, p. 500
6.	A	LO 7	Sizer/Whitney, p. 507
7.	B	LO 7	Sizer/Whitney, p. 509
8.	C	LO 7	Sizer/Whitney, p. 511
9.	A	LO 6	Video
10.	A	LO 5	Video
11.	A	LO 2	Video
12.	B	LO 7	Video
13.	C	LO 7	Video
14.	A	LO 7	Video
15.	C	LO 3	Video
16.	May include the following 6 to 8 wet diapers daily one bowel movement daily happy and alert baby appropriate weight gain	LO 5	Video
17.	A	LO 5	Sizer/Whitney, p. 503
18.	C	LO 7	Sizer/Whitney, p. 509
19.	E	LO 6	Sizer/Whitney, p. 510
20.		LO 6	Sizer/Whitney, p. 511
21.		LO 5	Sizer/Whitney, p. 503

Lesson 20

Life Cycle: Childhood and Adolescence

OVERVIEW

Childhood and adolescence are two periods during the life cycle that have their own unique nutrition challenges. In this lesson, you will consider the changes that take place during childhood and adolescence and how nutrition affects those changes. You will begin to see how a child's appetite and nutrient needs reflect the different stages of growth throughout childhood; how poor nutrition affects growth, learning, and behavior; and how parents can influence children's eating habits and lifestyles for the rest of their lives. You will discover why nutrient requirements increase dramatically during adolescence along with rapid growth spurts, and that teenagers have certain challenges they did not face as children, such as the strong influence of peers on lifestyle and nutrition habits. Finally, you will examine ways parents can help teenagers make wise choices about nutrition and lifestyles even when parents are no longer the focal point of their adolescents' lives.

LEARNING OBJECTIVES

Upon completing this lesson, you should be able to:

1. In general terms, describe the nutrient requirements for children.
2. Describe appropriate feeding practices for young children.
3. Describe how malnutrition affects growth, development, and behavior.
4. Distinguish between food allergy, intolerance, and aversion.
5. In general terms, describe the nutrient needs of adolescents.
6. Discuss the special nutrition concerns of teenagers.
7. Discuss the impact parents have in influencing their children's eating habits.
8. Explain how media influence adolescent behavior with regard to body image.

LESSON ASSIGNMENT

Review the following assignment in order to schedule your time appropriately. Pay careful attention; the titles and numbers of the textbook chapter, the telecourse guide lesson, and the video program may be different from one another.

Text: Sizer and Whitney, *Nutrition: Concepts and Controversies*, Chapter 14, "Child, Teen, and Older Adult," pp. 519–558

Video: "Life Cycle: Childhood and Adolescence" from the series *Nutrition Pathways*

LESSON FOCUS POINTS

The following questions are designed to help you get the most out of the lesson. After you read the textbook assignments and view the video, you should be able to answer these questions.

1. How many calories does a one-year-old need? A three-year-old? A ten-year-old? Why must parents limit access to concentrated sweets in a child's diet? (text, p. 522)

2. What is the most common nutrient deficiency in children in the United States and Canada? What is the effect on a child of having breakfast? What are the immediate and future effects of iron deficiency? What are general symptoms of a child suffering from nutrient deficiencies? How does lead poisoning affect a child's development? What is hyperactivity or attention-deficit/hyperactivity disorder (ADHD)? Does sugar cause hyperactivity? (text, pp. 522, 525–532)

3. Why should parents encourage children to help plan and prepare meals? What should parents do when a child won't accept new foods? When should new foods be introduced? What should parents do regarding snacking? What is the single most important influence on a child's eating habits? (text, pp. 524–525)

4. How can parents teach children about nutrition and encourage responsibility toward food? (text, p. 524; video)

5. What is the difference between food intolerance and food allergies? What three foods cause the majority of allergic reactions? What percent of food allergies in children can be attributed to peanuts, eggs, and milk? (text, pp. 527–529)

6. What impact can parents' habitual overeating have on children? (text, pp. 530–531; video)

7. At what ages do males and females go through adolescent growth spurts? Why do boys require more calories per day than girls of the same age? Why do both sexes require more iron and calcium at the onset of adolescence? (text, pp. 533–535)

8. What fraction of a teen's daily food intake is typically from snacks? What can parents do to ensure healthful snacking? What nutrients are generally lacking from fast foods that need to be available in other meals? Does chocolate or greasy food make acne worse? Will vitamin supplements improve acne? (text, pp. 536–537)

9. What is the first concrete element that passes between a primary caregiver and an infant? How does the importance of food impact future eating behavior? What can happen to children when parents do not set limits regarding food intake? (video)

10. How do programs, such as ShapeDown, help adolescents lose weight? What factor(s) most influence weight change in adolescents? (video)

11. How do media expectations regarding body image impact self-image and behavior in adolescent girls? (video)

12. How can food choices as an adolescent influence future development of type 2 diabetes? (video)

KEY CONCEPTS

Supervised Food Activities for Preschoolers

Children's muscle development determines their abilities to perform activities involving foods. The ages listed here are average ages; individual children develop on their own schedules.

A two-year-old can use large muscles of the arms to	▪ Scrub vegetables ▪ Tear lettuce ▪ Snap beans
A three-year-old can use hand muscles to	▪ Wrap foods in foil; wrap cheese slices around bread sticks ▪ Pour beverages from a small pitcher into a cup ▪ Mix cereal snacks in a large bowl with clean hands ▪ Shake juices together in a small sealed container ▪ Spread peanut butter, cream cheese, or jelly on bread with a dull butter knife or small spatula
A four-year-old can use the small muscles in the fingers to	▪ Peel the shells from hard-boiled eggs or the skins from oranges ▪ Roll out a ball of ground meat or dough with clean hands ▪ Juice citrus fruit by pushing down and turning the fruit on a hand juicer ▪ Crack raw eggs by tapping the center of the egg against the side of a bowl ▪ Mash a bowl of cooked beans for dip
A five-year-old can use eye-hand coordination to	▪ Measure ingredients with measuring spoons and cups ▪ Cut semisoft foods such as soft cheese with a dull butter knife and practice keeping fingers away from the blade ▪ Grind chunky peanut butter or cooked meats in a hand-turned food grinder, or turn a hand-cranked ice cream maker. ▪ Grate carrots or cheese on an upright grater with fingers far from the grater to avoid cuts

Source: Adapted from A. A. Hertzler, Preschoolers' food handling skills-motor development, Journal of Nutrition Education 21 (1989): 100B-100C.

PRACTICE TEST

The following items will help you evaluate your understanding of this lesson. Use the answer key at the end of the lesson to check your answers or to locate material related in each question.

Multiple Choice

Select the one choice that best answers the question.

1. A three-year-old child needs approximately ______ calories a day.
   A. 850
   B. 1,000
   C. 1,300
   D. 2,000

2. Which of the following is NOT a cause of "hyper" behavior in children?
   A. Lack of exercise
   B. Overstimulation
   C. Too much television
   D. Too much sugar

3. Which of the following describes foods preferred by children?
   A. Hot vegetables
   B. Mild flavors
   C. Overcooked vegetables
   D. Foods with lumps in them

4. An adverse reaction to food that does NOT involve the immune system is termed a __________.
   A. food allergy
   B. food intolerance
   C. mild food challenge
   D. food craving

5. Even in preschoolers whose habits are being established, existing dietary attitudes are relatively resistant to change. How should wise parents react?
   A. Suggest, rather than command, when offering a new food
   B. Impose their own eating habits on the children
   C. Wait until the children start school to initiate changes
   D. Constantly pressure the child to make good food choices

6. The adolescent growth spurt __________.
   A. primarily affects the brain
   B. decreases total nutrient needs
   C. affects every organ except the brain
   D. begins and ends earlier in girls than in boys

7. Approximately what fraction of an average teenager's daily energy intake is derived from snacks?
   A. One-fourth
   B. One-third
   C. One-half
   D. Two-thirds

8. Which of the following two conditions are associated with television's influence?
   A. Anorexia and nutrient deficiencies
   B. Obesity and high blood cholesterol
   C. Drug abuse and teenage pregnancy
   D. Hyperactivity and lower body weight

9. When food is overly important or not attended to properly in the family, experts have found that there __________.
   A. are more eating disorders
   B. are fewer incidences of binging and purging
   C. is a greater chance for discipline problems
   D. all of the above

10. Parents can teach children to become responsible with regard to food intake by __________.

A. modeling the behavior they want the children to learn
B. creating positive learning experiences around food
C. giving children a certain amount of authority with regard to food choices
D. all of the above

11. Children will learn to disregard their own internal hunger and satiety cues if they __________.

A. observe their parents habitually overeating
B. watch too much television, especially food advertisements
C. eat too many complex carbohydrates during the day
D. drink too many soft drinks

12. The weight loss program, ShapeDown, helps adolescents lose excess weight by __________.

A. treating not only the body but also the mind
B. utilizing an interdisciplinary team of professionals
C. involving all family members in the program
D. all of the above

13. Adolescent girls who do not measure up to media portrayal of the perfect young body image __________.

A. may develop eating disorders
B. develop a stronger self-image and more confidence
C. usually are more accepting of their bodies in spite of the magazines
D. frequently commit suicide

14. Adolescents who snack on high-fat foods or sugary foods will develop type 2 diabetes as adults if they __________.

A. are genetically predisposed
B. do not change their eating habits
C. gain too much weight
D. all of the above

Matching

Select the letter next to the word or phrase that best matches the numbered statements. Use answers only once.

_____	15. Environmental contaminant not normally found in food that is toxic when ingested	A. Lead
		B. Accutane
_____	16. Period in life when an individual becomes physically capable of reproduction	C. Puberty
_____	17. Medicine made from Vitamin A to treat acne and is highly toxic	D. Adulthood

Essay Questions

Reflect on the items listed below; one or more may be included on the exam.

18. Give examples of how hunger and nutrient deficiencies affect behavior in children.

19. Discuss the impact of television on children's eating behaviors and nutrition status.

ANSWER KEY

The following provides the answers and references for the practice test questions.

Answer		Learning Objectives	References
1.	C	LO 1	Sizer/Whitney, p. 522
2.	D	LO 3	Sizer/Whitney, p. 529
3.	B	LO 2	Sizer/Whitney, p. 524
4.	B	LO 4	Sizer/Whitney, p. 529
5.	A	LO 2	Sizer/Whitney, p. 524
6.	D	LO 5	Sizer/Whitney, p. 535
7.	A	LO 6	Sizer/Whitney, p. 536
8.	B	LO 8	Sizer/Whitney, p. 530
9.	A	LO 7	Video
10.	D	LO 7	Video
11.	A	LO 7	Video
12.	D	LO 6, 7	Video
13.	A	LO 8	Video
14.	A	LO 6	Video
15.	A	LO 3	Sizer/Whitney, p. 526
16.	C	LO 5	Sizer/Whitney, p. 535
17.	B	LO 6	Sizer/Whitney, p. 536
18.		LO 3	Sizer/Whitney, pp. 525–526
19.		LO 8	Sizer/Whitney, pp. 530–531

Lesson 21

Life Cycle: Adulthood and Aging

OVERVIEW

Wise food choices made throughout the growth years and into adulthood can help a person meet the many physical, emotional, and mental challenges of day-to-day living. Appropriate food choices and lifestyle habits can also help us to achieve freedom from diseases such as cardiovascular disease, diabetes, and cancer. The focus of this lesson reflects how nutrition and lifestyle habits can support or weaken a person's quest for longevity and wellness throughout adulthood and into the later years. In addition, you will see how genetics and the aging process play important roles in the quality of one's life, what nutrients are needed as people age, and how older adults can solve problems associated with a limited budget and nutrition concerns.

LEARNING OBJECTIVES

Upon completing this lesson, you should be able to:

1. Describe the nutrient needs of older adults.
2. Identify the effects of aging on nutrient status.
3. Describe the factors that affect longevity.
4. Cite risk factors associated with malnutrition in older adults.
5. Explain how nutrition may affect the development of cataracts.
6. Explain how nutrition may affect the development of Alzheimer's disease.

LESSON ASSIGNMENT

Review the following assignment in order to schedule your time appropriately. Pay careful attention; the titles and numbers of the textbook chapter, the telecourse guide lesson, and the video program may be different from one another.

Text: Sizer and Whitney, *Nutrition: Concepts and Controversies*, Chapter 14, "Child, Teen, and Older Adults," pp. 519–558

Video: "Life Cycle: Adulthood and Aging" from the series *Nutrition Pathways*

LESSON FOCUS POINTS

The following questions are designed to help you get the most out of the lesson. After you read the textbook assignments and view the video, you should be able to answer these questions.

1. What nutrition habits are associated with longevity? What health habits, other than good nutrition, are associated with longevity? What are the benefits of exercise for older adults? What is the effect of strength training in the elderly? What are the recommendations for physical activity in older people? (text, pp. 538–540)

2. What type of exercise is recommended to prevent loss of muscle mass? What type of exercise is recommended to prevent the onset of osteoporosis? (text, pp. 538–540; video)

3. At what age does the RDA for energy begin to decrease? What vitamin is absorbed and stored better in the elderly? What vitamin has the greatest risk for deficiency in the elderly and why? How does iron deficiency in older adults compare to younger people? Generally speaking, what are the recommendations regarding calcium intake in older people? What mineral is commonly deficient in the elderly? (text, pp. 538, 542–544)

4. What are cataracts? What nutrients are associated with cataract development? Define *osteoarthritis* and *rheumatoid arthritis*. How does nutrition affect arthritis? (text, pp. 541–543)

5. Why is dehydration a problem for many older people? What happens to body composition as we age? The GI tract? (text, p. 543)

6. Under what circumstances might older adults need supplementation? How do dental problems affect the elderly? What happens to the senses of taste and smell in the elderly? What other changes may affect nutrient status of the elderly? What are the typical food preferences in the elderly? Cite suggestions that can help older single people solve problems associated with buying, storing, and preparing foods for one person. (text, pp. 543–550)

7. What is Alzheimer's disease? What are the nutrition limits to Alzheimer's disease? (text, pp. 545–547)

8. How does aging affect the need for high-energy foods? (video)

9. What is the basic idea scientists are trying to get across to people with regard to aging? (video)

10. What are the effects of alcohol on an older person? (video)

11. How does diabetes impact the aging process? (video)

PRACTICE TEST

The following items will help you evaluate your understanding of this lesson. Use the answer key at the end of the lesson to check your answers or to locate material related in each question.

Multiple Choice

Select the one choice that best answers the question.

1. Studies of adults show that longevity is related, in part, to all of the following EXCEPT __________.
   A. weight control
   B. regularity of meals
   C. short periods of sleep
   D. no or moderate alcohol intake

2. After the age of 50, the RDA for energy assumes a _____ percent reduction in energy output per decade.
   A. 5
   B. 12
   C. 15
   D. 20

3. Absorption of ________ appears to increase with aging.
   A. vitamin A
   B. vitamin D
   C. calcium
   D. iron

4. Older adults should drink ________ glasses of water a day.
   A. three to four
   B. four to six
   C. six to eight
   D. eight to nine

5. What nutrients appear to be protective against cataract formation?
   A. Iron and calcium
   B. Chromium and zinc
   C. Vitamin B_{12} and folate
   D. Vitamin C and vitamin E

6. Older adults should __________.
   A. consume six to eleven servings of breads, grains, or pasta per day
   B. limit the amount of fat in their diets
   C. consume more protein than younger adults
   D. A and B

7. Which of the following is a characteristic of senile dementia of the Alzheimer's type?
   A. It affects 50 percent of those over eighty years of age
   B. It is responsive to dietary choline supplementation
   C. It is associated with stability of brain nerve cell number
   D. It is associated with elevated brain aluminum concentrations in most subjects

8. Alcohol can affect an older adult __________.
   A. by increasing blood alcohol level quickly
   B. because they are more sensitive to alcohol's effects
   C. because they have less body water to dilute it
   D. all of the above

9. A person who has lived with diabetes for twenty to forty years __________.
   A. is more likely to have complications associated with the disease
   B. is not likely to suffer any more ill effects than a younger person with the disease
   C. should pay especially close attention to the level of glycemic control
   D. A and C

Fill in the Blank

Insert the correct word or words in the blank for each item.

10. As you age, there is less of a need for foods that are __________________.

11. As a person ages, there is a relative loss of ____________________.

Matching

Select the letter next to the word or phrase that best matches the numbered statements. Use answers only once.

_____	12. Dietary restriction of this extends lifespan	A. Zinc
_____	13. Absorption of this nutrient appears to increase with age	B. Sodium
		C. Protein
_____	14. A deficiency of this mineral is common in older adults	D. Energy
_____	15. Need for this nutrient remains the same for both older and younger adults	E. Vitamin A

Essay Questions

Reflect on the items listed below; one or more may be included on the exam.

16. Why do older adults face a greater risk of vitamin D deficiency than younger people?

17. Describe the theory that links low antioxidant vitamin intakes with cataracts.

ANSWER KEY

The following provides the answers and references for the practice test questions.

	Answer	Learning Objectives	References
1.	C	LO 3	Sizer/Whitney, pp. 538–539
2.	A	LO 1	Sizer/Whitney, p. 539
3.	A	LO 2	Sizer/Whitney, p. 542
4.	C	LO 1	Sizer/Whitney, p. 543
5.	D	LO 5	Sizer/Whitney, p. 542
6.	D	LO 1	Sizer/Whitney, pp. 541–542
7.	D	LO 6	Sizer/Whitney, p. 545
8.	D	LO 2	Video
9.	D	LO 3	Video
10.	high in calories (fat, energy)	LO 1	Video
11.	muscle mass (lean body mass, muscle strength)	LO 2	Video
12.	D	LO 3	Sizer/Whitney, pp. 538–539
13.	E	LO 2	Sizer/Whitney, p. 542
14.	A	LO 4	Sizer/Whitney, p. 543
15.	C	LO 2	Sizer/Whitney, p. 540
16.		LO 4	Sizer/Whitney, p. 542
17.		LO 5	Sizer/Whitney, p. 542

Lesson 22

Diet and Health: Cardiovascular Disease

OVERVIEW

In this lesson you will learn about the disease that kills more people each year than any other disease—cardiovascular disease (CVD). Terms under the umbrella of CVD are atherosclerosis, heart disease, hypertension, and stroke. You will learn the ways each develops, the risk factors associated with each type, and diet and lifestyle choices that can protect you from developing CVD or help you recover if you already have cardiovascular disease.

LEARNING OBJECTIVES

Upon completing this lesson, you should be able to:

1. Define the terms associated with cardiovascular disease (CVD).
2. Explain how atherosclerosis develops.
3. Describe the risk factors associated with CVD.
4. Cite the recommendations for reducing CVD.
5. Explain how hypertension develops.
6. Describe the risk factors associated with hypertension.
7. Cite the recommendations for reducing hypertension.

LESSON ASSIGNMENT

Review the following assignment in order to schedule your time appropriately. Pay careful attention; the titles and numbers of the textbook chapter, the telecourse guide lesson, and the video program may be different from one another.

Text: Sizer and Whitney, *Nutrition: Concepts and Controversies*, Chapter 11, "Diet and Health," pp. 393–431

Video: "Diet and Health: Cardiovascular Disease" from the series *Nutrition Pathways*

LESSON FOCUS POINTS

The following questions are designed to help you get the most out of the lesson. After you read the textbook assignments and view the video, you should be able to answer these questions.

1. Describe the risk factors associated with hypertension. What are the symptoms? Which of the risk factors for hypertension might be modified by diet? (text, pp. 397, 408–412)

2. How does atherosclerosis usually begin? By what age do most people have well-developed plaques? How does plaque affect the blood pressure? What is a heart attack? What is a stroke. (text, pp. 398–400)

3. Define the following terms: cardiovascular disease (CVD), coronary heart disease (CHD), atherosclerosis, plaque, and hypertension. What are the leading causes of CVD death for adults? (text, pp. 398–402)

4. List the major risk factors associated with heart disease. What risk factors cannot be modified? What risk factor may be the most influential on CVD? What is considered a desirable blood cholesterol level? (text, pp. 401–405)

5. What is the number one cause of heart disease? What are the controllable risk factors? How often should people exercise to prevent heart disease? When do women typically develop heart disease? What are recommended levels of cholesterol? How does being overweight impact heart disease? (text, pp. 401–405; video)

6. What are two recommendations for reducing CVD risk? On what do dietary recommendations to reduce risk for CHD focus? What advice is offered to lower LDL? How does weight loss impact heart disease risk? What is the current evidence regarding alcohol intake and CVD risk? How does exercise impact CHD risk and cholesterol? How does exercise affect HDL? (text, pp. 405–408)

7. What percent of the adult population has hypertension? What are the symptoms? How does obesity affect blood pressure? How does insulin resistance impact blood pressure? What is considered normal blood pressure for adults? What blood pressure reading indicates hypertension? (text, pp. 408–411)

8. How does body fat distribution affect blood pressure? How does activity impact blood pressure? What does research state regarding salt/sodium intake and high blood pressure? How do alcohol and other dietary factors impact hypertension? How does weight loss affect hypertension? What is the role of calcium, potassium, and magnesium in regulating blood pressure? (text, pp. 411–412)

9. What is the recommended blood pressure? What are the recommendations to reduce hypertension? What is DASH? What is the recommended intake for sodium? (video)

KEY CONCEPTS

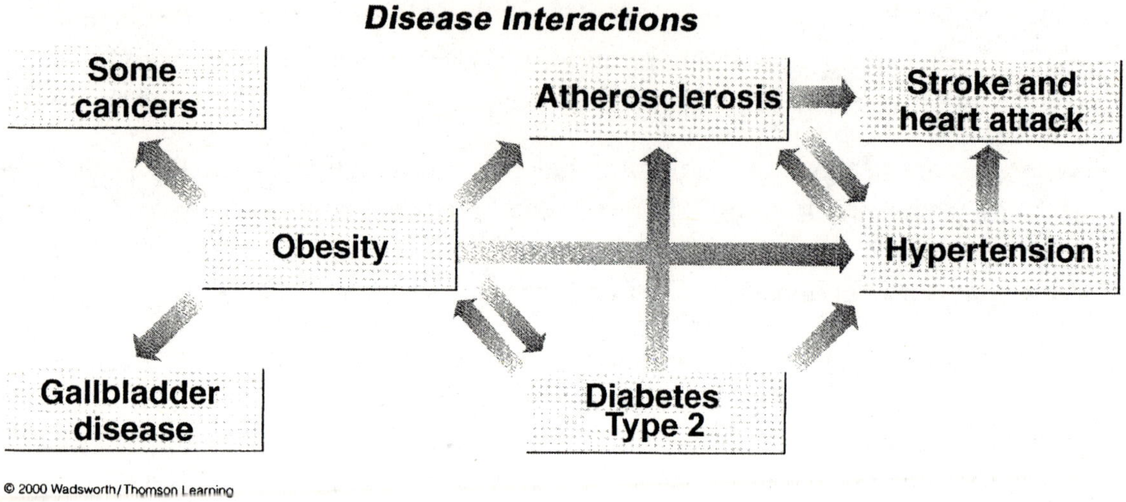

Major CV Disease Risk Factors

- Smoking
- High blood pressure
- Obesity (central or abdominal in particular)
- Elevated LDL cholesterol (blood)
- Lack of exercise

Other Risk Factors

- Low HDL cholesterol
- Family history
- Diabetes
- Male more than forty-five years of age
- Female after menopause

Recommendations for Preventing Hypertension

- Maintain desirable weight
- Increase physical activity
- Decrease alcohol use
- Lower sodium intake (for salt sensitive)
- Increase potassium intake
- Increase calcium intake
- Increase magnesium intake
- Increase vitamin C intake

RELATED ACTIVITY

This activity is not required unless your instructor assigns it. It is offered as a suggestion to help you learn more about the material presented in this lesson.

Complete the attached form for the following related activity and return it to your instructor according to the established deadline.

❑ How to Assess Your Heart Disease Risk

How to Assess Your Heart Disease Risk

Do you know your heart disease risk score? Respond to the statements below, and score yourself as directed. Be aware that a high risk score does not mean you will develop heart disease, but it should warn you of the possibility. Consult your physician if you have questions about your score results.

In each category, circle the number next to the statement that is most true for you.

Cigarette Smoking

- I never smoked or stopped smoking three or more years ago. 1
- I don't smoke but live and/or work with smokers. 2
- I stopped smoking within the past three years. 3
- I smoke regularly. 4
- I smoke regularly and live and/or work with other smokers. 5

Total Blood Cholesterol or

Use the number from your most recent blood cholesterol measurement:

- Less than 160 1
- 160-199 2
- Don't know 3
- 200-239 4
- 240 or higher 5

HDL Cholesterol

Use the number from your most recent HDL cholesterol measurement:

- Over 60 1
- 56-60 2
- Don't know 3
- 35-55 4
- Less than 35 5

Systolic Blood Pressure

Use the first (highest) number from your most recent blood pressure measurement:

- Less than 120 1
- 120-139 2
- Don't know 3
- 140-159 4
- 160 or higher 5

Excess Body Weight

- I am within 10 pounds of my desirable weight 1
- I am 10-20 pounds above my desirable weight 2
- I am 21-30 pounds above my desirable weight 3
- I am 31-50 pounds above my desirable weight 4
- I am more than 50 pounds above my desirable weight 5

Source: American Heart Association

Physical Activity

Determine which statements best describe your usual level of physical activity:

A. Highly Active
- My job requires very hard physical labor (such as digging or loading heavy objects) at least four hours a day or
- I do vigorous activities (jogging, cycling, swimming, etc.) at least three times a week for 30 minutes or more or
- I do at least one hour of moderate activity such as brisk walking at least four days a week.

B. Moderately Active
- My job requires that I walk, lift, carry, or do other moderately hard work several hours a day (day-care worker, stock clerk, or busboy/waitress) or
- I spend much of my leisure time doing moderate activities (dancing, gardening, walking, or housework).

C. Inactive
- My job requires that I sit at a desk most of the day and
- Much of my leisure time is spent in sedentary activities (watching TV, reading, etc.) and
- I seldom work up a sweat, and I cannot walk fast without having to stop to catch my breath.

Circle the number that best describes your level of physical activity.

A:	Highly Active	1
	Between A and B	2
B:	Moderately Active	3
	Between B and C	4
C:	Inactive	5

Scoring Your Heart Attack Risk

To learn your estimated risk, add the six numbers you've circled.

If your Total Score Is:	***Your Heart Attack Risk Is:***
6-13	Low
14-22	Moderate
23-30	High

PRACTICE TEST

The following items will help you evaluate your understanding of this lesson. Use the answer key at the end of the lesson to check your answers or to locate material related in each question.

Multiple Choice

Select the one choice that best answers the question.

1. What is the term given to mounds of lipid material mixed with smooth muscle cells and calcium that develop in the artery walls?
   A. Plaques
   B. Atherosclerosis
   C. Arterial thickening
   D. Cardiovascular disease

2. Most people have well-developed plaques by the time they reach age _______.
   A. 15
   B. 25
   C. 30
   D. 45

3. Wherever dietary fat consists mostly of _____ fats, and fish, fruit, and vegetables are in eaten in abundance, blood cholesterol and the rate of death from heart disease are low.
   A. saturated
   B. polyunsaturated
   C. monounsaturated
   D. omega-6

4. What type of exercise is best to strengthen the heart and blood vessels?
   A. Aerobic exercise
   B. Anaerobic exercise
   C. Heavy weight training
   D. Short duration exercise

5. Salt-sensitive individuals are likely to include __________.
   A. African Americans
   B. those who are less than 50 years old
   C. those who have diabetes
   D. A and C

6. Which of the following statements describes a relationship between salt intake and high blood pressure?
   A. Most people can safely ingest more salt than needed because of efficient salt regulation mechanisms.
   B. Ninety-five percent of people under 50 years of age with high blood pressure respond favorably to low salt intake.
   C. Most people are genetically sensitive to sodium and can develop high blood pressure from excess salt intake.
   D. Ninety five percent of people with high blood pressure who restrict salt intake show a significant decrease in blood pressure.

7. Which of the following appear to help prevent and treat hypertension in certain populations?
   A. Calcium
   B. Potassium
   C. Folate
   D. A and B

8. Which of the following increase the risk for developing high blood pressure?
   A. Obesity
   B. Smoking
   C. Lack of exercise
   D. All of the above

9. On the average and compared to men, women tend to get heart disease ______.
   A. as frequently as men
   B. about ten to fifteen years later than men
   C. but have fewer complications than men
   D. all of the above

10. Populations that may be more susceptible to potential heart damage caused by elevated triglycerides include __________.
    A. younger men
    B. people with diabetes
    C. people with cancer
    D. A and B

11. When LDL level is elevated and HDL is lower than recommended, the first thing that should be examined is a person's __________.
    A. alcohol intake
    B. level of exercise
    C. genetic background
    D. smoking habits

12. With respect to vitamin supplementation for reducing heart disease risk, experts recommend __________.
    A. that men do not take iron supplements
    B. modest doses of antioxidants, if at all
    C. obtaining vitamins from whole foods
    D. all of the above

Matching

Select the letter next to the word or phrase that best matches the numbered statements. Use answers only once.

_____	13. Also known as high blood pressure	A. Hypertension
_____	14. A mound of lipid material embedded in arterial walls	B. Plaque
		C. Heart attack
_____	15. Results from a clot that stops blood flow to the heart	D. Stroke
_____	16. Results from a clot that stops blood flow to the brain	E. Cardiovascular disease

Essay Questions

Reflect on the items listed below; one or more may be included on the exam.

17. Describe dietary and other strategies for reducing risk of CVD.

18. Explain how atherosclerosis can make hypertension worse.

ANSWER KEY

The following provides the answers and references for the practice test questions.

	Answer	Learning Objectives	References
1.	A	LO 1	Sizer/Whitney, p. 399
2.	C	LO 1	Sizer/Whitney, p. 399
3.	C	LO 3	Sizer/Whitney, p. 406
4.	A	LO 4	Sizer/Whitney, p. 404
5.	D	LO 6	Sizer/Whitney, p. 411
6.	A	LO 6, 7	Sizer/Whitney, p. 411
7.	D	LO 7	Sizer/Whitney, pp. 411–412
8.	D	LO 6, 5	Video
9.	B	LO 3	Video
10.	D	LO 3	Video
11.	B	LO 3	Video
12.	D	LO 4	Video
13.	A	LO 1	Sizer/Whitney, p. 397
14.	B	LO 1	Sizer/Whitney, p. 399
15.	C	LO 2	Sizer/Whitney, p. 400
16.	D	LO 2	Sizer/Whitney, p. 400
17.		LO 4	Sizer/Whitney, pp. 405–407
18.		LO 2	Sizer/Whitney, p. 400

Lesson 23

Diet and Health: Cancer, Immunology, and AIDS

OVERVIEW

Cancer—the most dreaded of all diseases—is the second leading cause of death in the United States. Some cancers can be cured if detected in time; others are incurable; still others can be prevented altogether if proper changes in lifestyle and eating habits are made in a timely manner. This lesson provides you with the information necessary to take action against the onset of cancer and includes an explanation of how cancer develops, the risk factors associated with cancer, and the recommendations for reducing the risk of cancer. In addition to learning about cancer, you will also learn about the basic functioning of the immune system, the impact of nutrition and lifestyle on immunity, and how nutrition intervention can help alleviate the symptoms and improve the outcome of HIV/AIDS (human immunodeficiency virus/acquired immune deficiency syndrome) patients.

LEARNING OBJECTIVES

Upon completing this lesson, you should be able to:

1. Describe the process by which a cancer develops.
2. Describe how diet may play a role in the development of cancer.
3. Cite the recommendations for reducing cancer risk.
4. Describe the impact of nutrition on the immune system.
5. Define HIV and AIDS.
6. Explain how nutrition intervention can support people with HIV/AIDS.

LESSON ASSIGNMENT

Review the following assignment in order to schedule your time appropriately. Pay careful attention; the titles and numbers of the textbook chapter, the telecourse guide lesson, and the video program may be different from one another.

Text: Sizer and Whitney, *Nutrition: Concepts and Controversies*, Chapter 11, "Diet and Health," pp. 393–431

Video: "Diet and Health: Cancer, Immunology, and AIDS" from the series *Nutrition Pathways*

LESSON FOCUS POINTS

The following questions are designed to help you get the most out of the lesson. After you read the textbook assignments and view the video, you should be able to answer these questions.

1. Define *immune system*. List four ways nutrition interacts with the immune system. What groups of people are likely to be caught in the downward spiral of malnutrition and infection? (text, pp. 394–396)

2. List three ways adequate nutrition can help a person with HIV/AIDS. (text, pp. 395–396)

3. Define HIV and AIDS. What impact does malnutrition have on AIDS? What nutrition strategies are provided to AIDS patients? Describe how nontraditional supplements might impact AIDS. What are the dietary guidelines for people with AIDS? What are the most commonly malabsorbed nutrients in AIDS patients? What is the recommendation concerning vitamin supplements for AIDS patients? Describe the impact of anorexia on AIDS patients. (text, p. 396; video)

4. Cite the guidelines for reducing the risks for cancers. (text, p. 413)

5. List the six steps of how cancer is thought to develop. (text, p. 416)

6. How do genetics influence the development of cancer? What are environmental factors that are known to cause cancer? What types of cancers

are related to dietary factors? How do dietary fats impact cancer? What dietary factors promote or may protect against cancer? (text, pp. 416–420)

7. What is the impact of fat on the development of breast cancer? What is the recommended fat intake to prevent breast cancer? What are the six suggestions from the American Cancer Society regarding cancer prevention? Which type of fiber has an impact on cancer development? (video)

8. How does immune function affect nutritional status of hospitalized people? Why do experts study the elderly with regard to immune function? What are simple rules to follow regarding strengthening the immune system? What do experts advise with regard to antioxidant supplementation and immune system function and health? (video)

KEY CONCEPTS

Stages in Cancer Development

I. Exposure to a carcinogen
II. Carcinogen enters cell
III. Initiation—alteration of DNA
IV. Promotion—development of abnormal cells
V. Tumor formation
VI. Disruption of normal body functions

Nutrition and Immunity

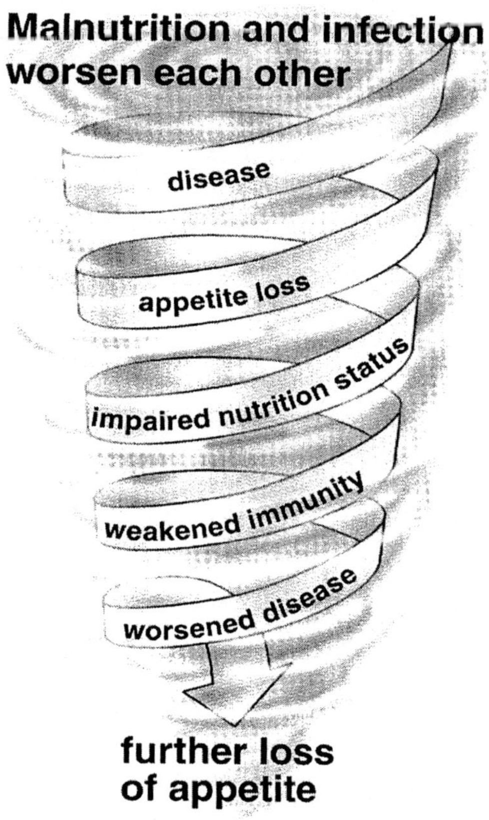

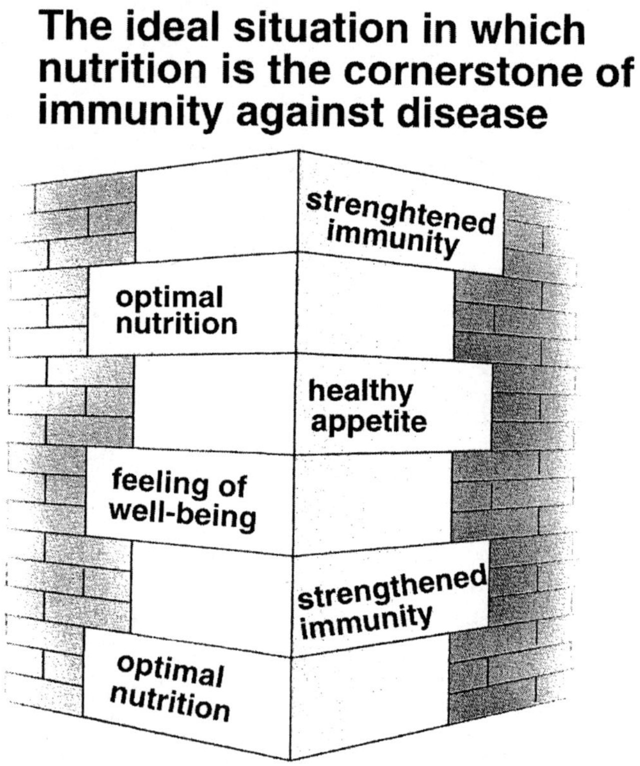

© 2000 Wadsworth/Thomson Learning

Possible Dietary Factors in the Development of Cancer

I. Initiators
- Pesticides
- Nitrosamines
- Alcohol

II. Promoters
- Fat intake, especially omega-6 fatty acid, linoleic acid

III. Possible Preventers
- Fiber
- Antioxidant nutrients
- Phytochemicals (soy)
- Cruciferous vegetables
- Omega-3 fatty acids

Diet Recommendations for Cancer Prevention

- Maintain desirable weight
- Fat intake 30 percent or less of total calories (saturated fat less than 10 percent) to increase fiber intake (20 to 30 grams)
- Increase intake of fruits and fresh vegetables, as they contain antioxidants and phytochemicals
- Avoid smoked and pickled foods, as they contain nitrates
- Use alcohol in moderation, if at all
- Limit intake of acrylamide

PRACTICE TEST

The following items will help you evaluate your understanding of this lesson. Use the answer key at the end of the lesson to check your answers or to locate material related in each question.

Multiple Choice

Select the one choice that best answers the question.

1. Which of the following holds the key to maintaining the best possible immune system support?
   A. Nutrient supplements
   B. Herbal remedies
   C. Phytochemicals
   D. A well-balanced diet

2. Which of the following must be given careful attention when feeding a person with AIDS?
   A. Soft foods
   B. Small frequent snacks
   C. Food safety
   D. Daily Food Guide recommendation

3. Antioxidant substances that are believed to help protect against cancer include all of the following EXCEPT __________.
   A. folate
   B. vitamin E
   C. vitamin C
   D. beta-carotene

4. Which of the following is NOT among the recommendations issued by health professionals to reduce cancer risks?
   A. Avoid obesity
   B. Moderate or stop intake of alcohol
   C. Increase intake of foods high in iron
   D. Limit intake of foods preserved with smoke or salts

5. In general, studies of populations have suggested that low cancer rates correlate with __________.
   A. high vegetable intake
   B. high grain intake
   C. high fat intake
   D. A and B

6. The best action to take to decrease your risk of cancer is to __________.
   A. eat fruits and vegetables daily
   B. consume a diet moderate in fat
   C. take vitamin supplements
   D. eat a wide variety of vegetables and fruits in generous quantities every day

7. Experts suggest that the threshold of fat intake to prevent breast cancer is __________ of total calories.
   A. less than 30 percent
   B. less than 20 percent
   C. about 30 percent
   D. about 45 percent

8. The American Cancer Society suggests that women and men maintain desirable weight based on the following criteria:
   A. women should weigh 100 pounds for 5 feet of height plus 5 pounds for each inch over 5 feet.
   B. both women and men should weigh 106 pounds for 5 feet of height plus 6 pounds for each inch over 5 feet.
   C. men should weigh 106 pounds for 5 feet of height plus 6 pounds for each inch over 5 feet.
   D. A and C.

9. Among hospitalized patients, health professionals see a general pattern with regard to immune function in that __________.
   A. appetite decreases and patients become malnourished
   B. immune function improves because of overall good care
   C. lack of nutrients leads to faulty immune function
   D. A and C

10. By studying the immune systems of elderly people, health professionals are learning __________.
   A. what nutrients appear promising in reducing risk of diseases such as cancer
   B. that nutrition helps people grow old gracefully
   C. to turn back the hands of time by supplementing with specific nutrients
   D. all of the above

11. HIV can be defined as __________.
   A. a virus
   B. an infection that attacks the immune system
   C. a disease that attacks the T-4 cells in the body
   D. all of the above

12. The most commonly malabsorbed nutrient(s) in AIDS patients is (are) ______.
   A. lactose or milk sugar
   B. protein or amino acids
   C. fatty acids
   D. A and C

13. Very high protein intakes are recommended to AIDS patients because they __________.
   A. become anorexic and lose lean body mass
   B. are in negative nitrogen balance due to wasting
   C. cannot tolerate high carbohydrate foods as a rule
   D. A and B

Matching

Select the letter next to the word or phrase that best matches the numbered statements. Use answers only once.

______	14. A substance that causes alteration of DNA	A. Promoter
______	15. A substance that allows development of a cancer after the DNA has been altered	B. Inhibitor
		C. Initiator

Essay Questions

Reflect on the items listed below; one or more may be included on the exam.

16. Outline the steps involved in the development of cancer.

17. How does AIDS develop?

ANSWER KEY

The following provides the answers and references for the practice test questions.

	Answer	Learning Objectives	References
1.	D	LO 4	Sizer/Whitney, p. 394
2.	C	LO 6	Sizer/Whitney, p. 396
3.	A	LO 2	Sizer/Whitney, pp. 419–420
4.	C	LO 3	Sizer/Whitney, p. 413
5.	D	LO 2	Sizer/Whitney, p. 419
6.	D	LO 3	Sizer/Whitney, pp. 419–420
7.	B	LO 3	Video
8.	D	LO 3	Video
9.	D	LO 4	Video
10.	A	LO 4	Video
11.	D	LO 5	Video
12.	D	LO 6	Video
13.	D	LO 6	Video
14.	C	LO 1	Sizer/Whitney, p. 416
15.	A	LO 1	Sizer/Whitney, p. 416
16.		LO 1	Sizer/Whitney, p. 416
17.		LO 5	Sizer/Whitney, pp. 395–396

Lesson 24

Diet and Health: Diabetes

OVERVIEW

The seventh leading cause of death in the United States is diabetes. Not only is it among the top ten causes of death, but people with diabetes are also two times as likely to die from cardiovascular conditions such as heart attack and stroke as people without the disease. In this lesson you will explore the differences between the two most common forms of diabetes mellitus—insulin-dependent diabetes mellitus (IDDM) and noninsulin-dependent diabetes mellitus (NIDDM). Specifically, you will learn which form is most common, what complications are associated with each form, what the dietary recommendations are for IDDM and NIDDM, and how exercise impacts the diabetic condition.

LEARNING OBJECTIVES

Upon completing this lesson, you should be able to:

1. Compare and contrast IDDM and NIDDM.
2. Explain the complications associated with diabetes.
3. Compare and contrast the dietary recommendations for IDDM and NIDDM.
4. Explain the effects of exercise on diabetes.
5. Explain the effects of nutrition therapy on diabetes.

LESSON ASSIGNMENT

Review the following assignment in order to schedule your time appropriately. Pay careful attention; the titles and numbers of the textbook chapter, the telecourse guide lesson, and the video program may be different from one another.

Text: Sizer and Whitney, *Nutrition: Concepts and Controversies*, Chapter 4, "The Carbohydrates: Sugar, Starch, Glycogen, and Fiber," pp. 99–138

Video: "Diet and Health: Diabetes" from the series *Nutrition Pathways*

LESSON FOCUS POINTS

The following questions are designed to help you get the most out of the lesson. After you read the textbook assignments and view the video, you should be able to answer these questions.

1. Define the following terms: insulin-dependent (type 1, juvenile-onset) diabetes mellitus (IDDM); noninsulin-dependent (type 2, adult-onset) diabetes mellitus (NIDDM); and insulin resistance. (text, pp. 121–122)

2. What is the focus of nutrition therapy in diabetes? Under what three circumstances must a person with type 1 diabetes adjust insulin dosage? What is the effect of moderate weight loss on type 2 diabetes? (text, pp. 121–122)

3. Why must people with type 1 diabetes take insulin injections? Describe the acute symptoms associated with type 1 diabetes. (text, pp. 121–122; video)

4. What percent of the population in the United States has IDDM (type 1, juvenile-onset)? What is the function of insulin? Why must insulin be injected rather than taken orally? What percent of the population in the United States has NIDDM (type 2, adult-onset)? In what population does NIDDM most often develop? What percent of U.S. adults with NIDDM are obese? (text, pp. 121–123)

5. What are the difficulties associated with diagnosis of type 2 diabetes? What population is the highest risk group for type 2 diabetes? (text, pp. 122–123)

6. Describe the complications associated with diabetes. Approximately what percent of people with diabetes dies from CVD? Approximately what percent of people with diabetes have kidney or vision impairment? What is gangrene and what can happen as a result of gangrene? (text, p. 123)

7. What is the impact of physical activity on insulin and blood glucose? (text, pp. 123–124)

8. Why is exercise encouraged early on in the management of type 2 diabetes? (text, pp. 123–124; video)

9. What is the primary role for diabetes educators with regard to parents of children with type 1 diabetes? Why should children with type 1 diabetes be monitored carefully? What dietary restrictions do children with type 1 diabetes have? What is the benefit of exercise for children with type 1 diabetes? (video)

10. What four factors are associated with the onset of type 2 diabetes? What is the number one condition associated with type 2 diabetes? What percent of nontraumatic, nonaccidental amputations is associated with type 2 diabetes? What percent of new cases of renal failure is attributed to type 2 diabetes? What percent of people with type 2 diabetes will die from cardiovascular disease? (video)

11. What are the effects of a raw vegetable diet on type 2 diabetes? (video)

KEY CONCEPTS

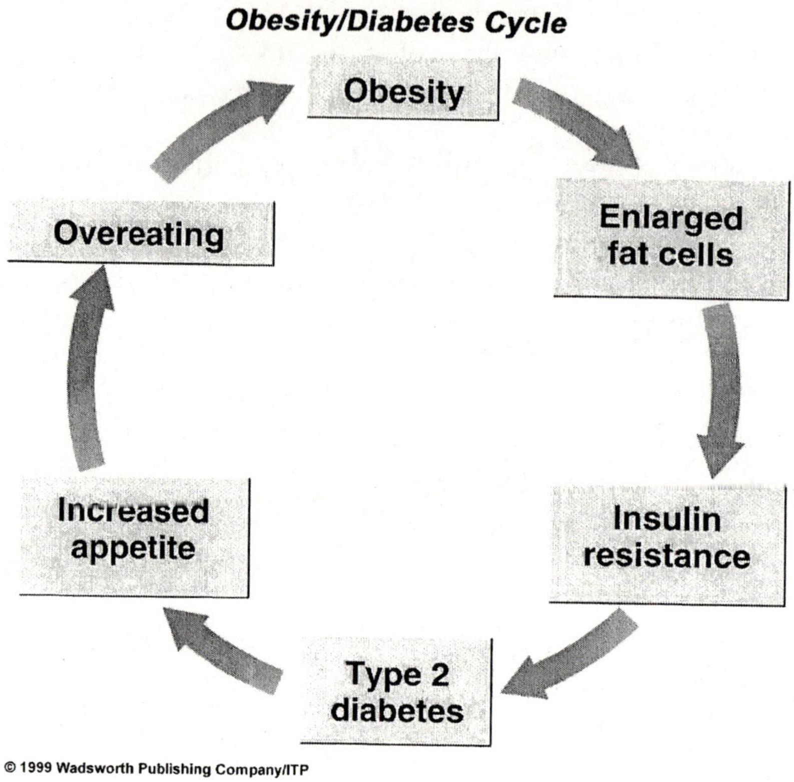

PRACTICE TEST

The following items will help you evaluate your understanding of this lesson. Use the answer key at the end of the lesson to check your answers or to locate material related in each question.

Multiple Choice

Select the one choice that best answers the question.

1. Characteristic(s) of type 1 diabetes include(s) __________.
   A. the person's immune system attacks the cells of the pancreas
   B. the pancreas no longer produces insulin
   C. the person is usually overweight
   D. A and B

2. In which of the following conditions would the pancreas be UNABLE to synthesize the hormone insulin?
   A. Hyperglycemia
   B. Type 1 diabetes mellitus
   C. Type 2 diabetes mellitus
   D. Adult-onset diabetes mellitus

3. A person with diabetes is most likely to develop __________.
   A. AIDS
   B. cancer
   C. diverticulosis
   D. strokes and heart attacks

4. Heavy use of _______ makes development of type 2 diabetes more likely.
   A. sweets
   B. fats
   C. alcohol
   D. protein

5. Insulin injections are required daily for people with type I diabetes because __________.
   A. their bodies do not produce insulin
   B. it prevents many metabolic problems from occurring
   C. insulin stimulates the cells to remove glucose from the blood
   D. all of the above

6. Because onset is more gradual with type 2 diabetes, making diagnosis difficult, experts surmise that more than __________ of people with diabetes remain undiagnosed.
   A. one-fourth
   B. one-third
   C. one-half
   D. three-fourths

7. The population at highest risk for type 2 diabetes is __________.
   A. males under forty years
   B. females over forty years
   C. children under eleven years
   D. adolescents between fourteen and eighteen years

8. The primary role of diabetes educators when children with type 1 diabetes are released from the hospital is to __________.
   A. teach parents/families how to give insulin injections
   B. give the children their insulin injections until they are old enough to give them to themselves
   C. teach parents how to perform blood glucose tests
   D. A and C

9. Care must be taken with regard to exercise in children with type 1 diabetes because exercise __________.
   A. can decrease the amount of insulin needed during the day
   B. has been shown to increase the amount of insulin needed
   C. is not beneficial to children with type 1 diabetes
   D. A and C

10. Of which of the following conditions/diseases is diabetes the number one cause?
    A. Heart disease
    B. Blindness
    C. Amputation
    D. Kidney failure

11. Glucose uptake can be increased by 10 to 20 percent above normal if a person with diabetes __________.
    A. takes oral agents in addition to insulin injections
    B. eats only foods high in complex carbohydrates and animal protein
    C. exercises on a regular basis
    D. all of the above

12. Exercise is encouraged early on in the treatment of diabetes because exercise can __________.
   A. cause muscles to take up ten to twenty times more glucose than nonexercising muscles
   B. prevent the heart problems associated with diabetes
   C. prevent amputations because exercising the legs stimulates the release of stored glycogen
   D. burn excess calories and reduce body fat associated with the onset of diabetes

Matching

Select the letter next to the word or phrase that best matches the numbered statements. Use answers only once.

______ 13. This type of diabetes is usually controlled without insulin injections

______ 14. One of the most common long-term complications of diabetes

A. Blindness

B. Frequent urination

C. Type 1 diabetes

D. Type 2 diabetes

Essay Questions

Reflect on the item listed below; it may be included on the exam.

15. Compare and contrast the two major types of diabetes and their recommended dietary management.

ANSWER KEY

The following provides the answers and references for the practice test questions.

	Answer	Learning Objectives	References
1.	D	LO 1	Sizer/Whitney, p. 122
2.	B	LO 1	Sizer/Whitney, p. 121
3.	D	LO 2	Sizer/Whitney, p. 123
4.	C	LO 1	Sizer/Whitney, p. 123
5.	D	LO 1	Video
6.	C	LO 1	Video
7.	B	LO 1	Video
8.	D	LO 1	Video
9.	A	LO 4	Video
10.	B	LO 2	Video
11.	C	LO 1, 2, 3	Video
12.	A	LO 4	Video
13.	D	LO 1	Sizer/Whitney, p. 123
14.	A	LO 2	Sizer/Whitney, p. 123
15.		LO 3	Sizer/Whitney, pp. 123–124

Lesson 25

Consumer Concerns and Food Safety

OVERVIEW

As consumers we are all concerned about what goes into the foods we eat and the water we drink before they get to our tables. How do pesticides and additives affect the foods we eat and our health? Should we worry about the public water supply and drink only bottled water? Specifically, this lesson will introduce you to food-borne illnesses and explain how you can prevent them by practicing food safety in your home and while traveling; how environmental contaminants make their way into the food and water supply; what the functions and hazards associated with pesticide use are; how food additives affect foods and consumer health; how the public water supply may be contaminated and how it is protected; and how bottled water issues affect consumers.

LEARNING OBJECTIVES

Upon completing this lesson, you should be able to:

1. List in order of priority the five hazards in the U.S. food supply that have been identified by the FDA.
2. Discuss food poisoning prevention including those foods identified as the most frequent sources of food poisoning.
3. Describe in general terms how environmental contaminants find their way into the food supply.
4. Distinguish the risks among the following: natural toxins, hormone residues, and pesticides.
5. Identify six types of food additives that are used in the food supply and how each type of food additive may affect health.
6. Compare and contrast the sources of tap water and bottled water.
7. Describe the safety and efficacy of irradiation.

LESSON ASSIGNMENT

Review the following assignment in order to schedule your time appropriately. Pay careful attention; the titles and numbers of the textbook chapter, the telecourse guide lesson, and the video program may be different from one another.

Text: Sizer and Whitney, *Nutrition: Concepts and Controversies*, Chapter 12, "Food Safety and Food Technology," pp. 433–480
Read this chapter in an overview fashion, concentrating only on the text and video focus points associated with the lesson.

Video: "Consumer Concerns and Food Safety" from the series, *Nutrition Pathways*

LESSON FOCUS POINTS

The following questions are designed to help you get the most out of the lesson. After you read the textbook assignments and view the video, you should be able to answer these questions.

1. What U.S. agency monitors the food supply? What are the possible hazards in our food supply? Which is of the most concern? Which is of least concern? (text, pp. 434–435)

2. What populations are most vulnerable to food-borne illnesses? What three steps can be taken to prevent food-borne illnesses from arising in the kitchen? What is a general rule concerning a suspect food? What foods are most likely to carry microbes that will cause food poisoning? (text, pp. 434–438)

3. What are the three components of safe food handling? What is cross-contamination? What advice is given when handling fruits and vegetables during preparation? (text, pp. 440–441; video)

4. Who is at risk for pesticide hazards? How can consumers minimize risks associated with pesticide use? What are alternatives to pesticide use? What are natural toxins? Can they be harmful? What is bovine somatotropin (BST)? Why is it used in cattle? Is it safe? (text, pp. 452–455)

5. What are the steps in the bioaccumulation of toxins in the food chain? Name two recent examples of environmental contamination. (text, pp. 456–459)

6. What is the general function of additives in the food supply? Explain how the following intentional food additives affect food and/or health: antimicrobial agents, antioxidants, artificial colors, flavor enhancers, nutrient additives. (text, pp. 459–467)

7. What are some of the side effects of monosodium glutamate (MSG) as a food additive? What can people do to avoid MSG? (text, pp. 466–467; video)

8. What are alternate methods of growing herbs without use of pesticides that also enhance the quality of the soil? What types of plants or insects may be used to repel harmful plant pests? (text, pp. 472–480; video)

9. What are the two sources of drinking water? What is the source of each? How can each source become contaminated? Why is the use of bottled water increasing? How is bottled water disinfected? (video)

10. What is a common belief regarding bottled water versus tap water? How do bottled water sales compare to other beverages? How can consumers determine the source of bottled water? (video)

Temperature of Food

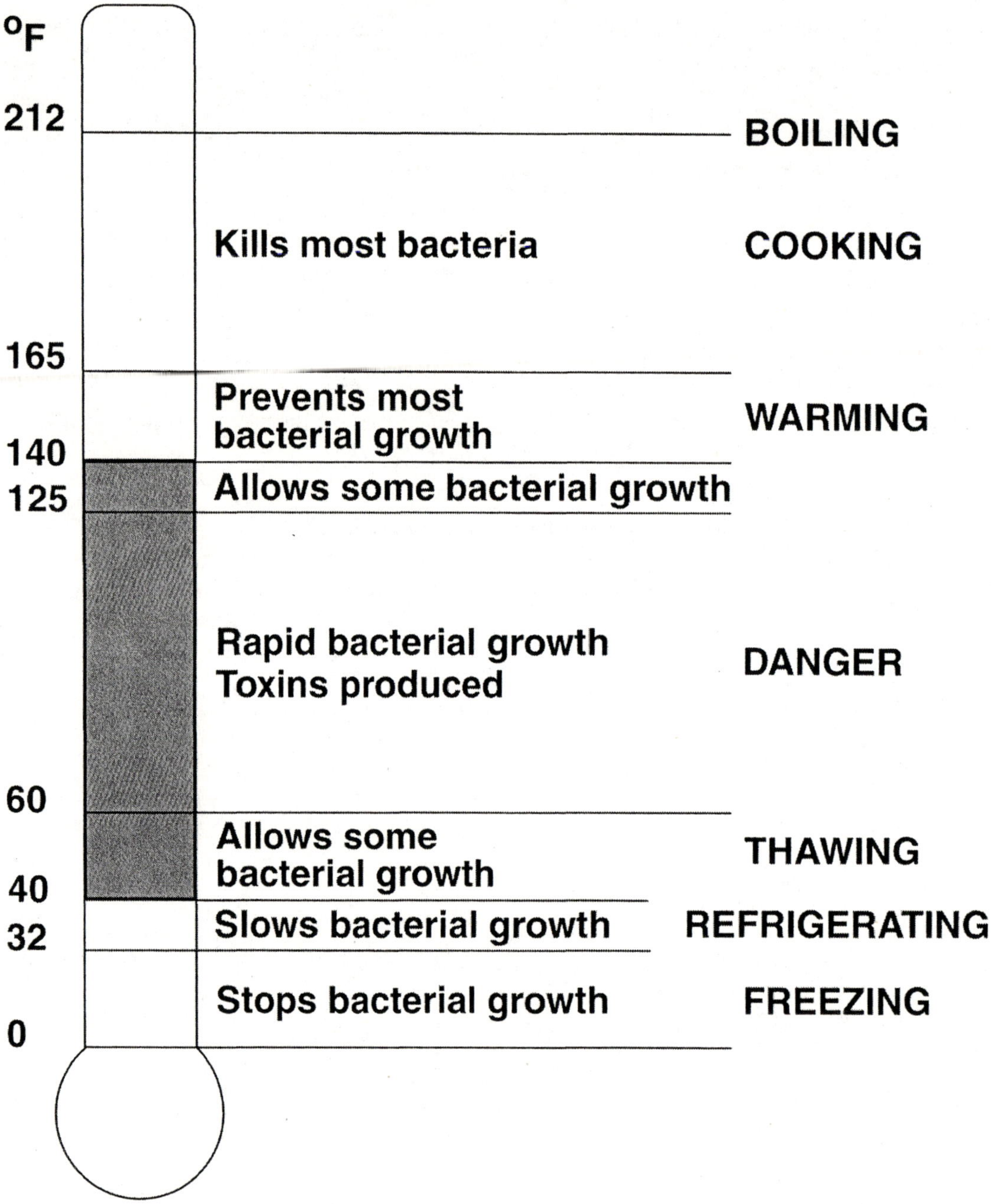

© 2000 Wadsworth/Thomson Learning

How To Prevent Food Poisoning: Clean Kitchen

- ❑ Use warm soapy water to wash hands, utensils, dishes, cutting boards, and countertops.
- ❑ Avoid cross-contamination by washing all surfaces that have been in contact with raw meats, poultry, or eggs before using.
- ❑ Mix food with utensils, not hands; keep hands and utensils away from mouth, nose, and hair.
- ❑ Avoid coughing or sneezing over food. A person with a skin infection or infectious disease should not prepare food.

Wash or replace sponges and towels regularly. Clean up food spills and crumb-filled crevices.

- ❑ Throw out foods with off odors.
- ❑ Do not even taste food that is suspect.
- ❑ Do not buy or use items that appear to have been opened; check safety seals, buttons, and rings. Observe expiration dates.
- ❑ Follow label instructions for storing and preparing packaged and frozen foods.

© 2000 Wadsworth/Thomson Learning

How to Prevent Food Poisoning: Keep Foods Hot

- ❑ Use a meat thermometer to test the internal temperature of meats and poultry.
- ❑ Insert the thermometer between the thigh and the body of the turkey or in the thickest part of other meats, making sure the tip of the thermometer is not in contact with the bone. Cook to the temperature indicated for that particular meat; cook hamburgers to at least medium well-done.
- ❑ Cook stuffing separately or stuff poultry just prior to cooking.

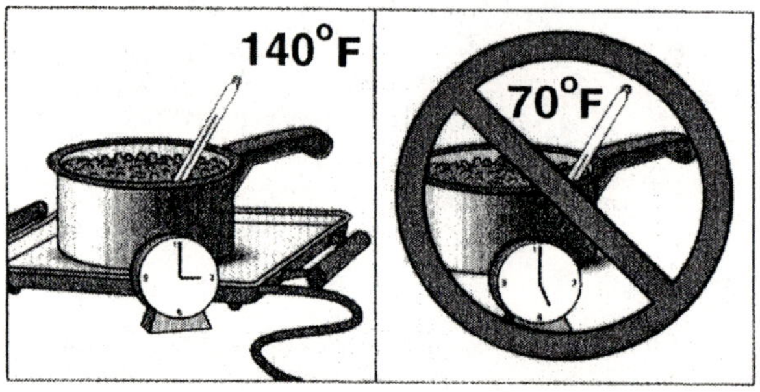

- ❑ When serving foods for longer than two hours, maintain temperatures about 140°F. Heat leftovers thoroughly to at least 140°F.

- ❑ Cook eggs and seafood thoroughly before eating them.

© 2000 Wadsworth/Thomson Learning

How to Prevent Food Poisoning: Keep Foods Cold

- ❑ Keep cold foods at least 40°F or less.
- ❑ Refrigerate leftovers promptly; use shallow containers to help foods cool faster.

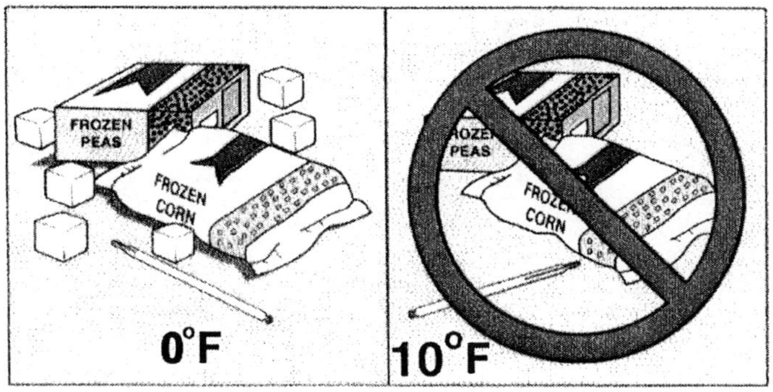

- ❑ Keep frozen foods at least 0°F or less.

- ❑ Tote lunches in a thermal bag or box.
- ❑ Freeze plastic bottles or pouches of beverages and let them keep the lunch cool as they thaw out through the morning.

© 2000 Wadsworth/Thomson Learning

How to Detect and Avoid Spoiled Foods

These Foods	Are Risky When
Fresh poultry	▪ Stored raw in the refrigerator for longer than 1-2 days, (3-4 days for cooked poultry) ▪ Left unrefrigerated for more than 2 hours either before or after cooking
Fresh meat	▪ Stored raw in the refrigerator for longer than 3-5 days (1-2 days for hamburger) ▪ Discolored, smelling, slimy ▪ Left unrefrigerated for more than 2 hours either before or after cooking.
Fresh fish	▪ Stored for longer than 1-2 days in the refrigerator ▪ Dried at edges, smelly ▪ Left unrefrigerated for more than 2 hours whether before or after cooking
Milk, cream, egg products	▪ Left unrefrigerated for more than 2 hours ▪ Stored in the refrigerator longer than 5-7 days
Frozen meats, poultry, fish, casseroles	▪ Thawed at room temperature ▪ Allowed to thaw and be refrozen ▪ Eaten without thorough cooking
Canned foods*	▪ Liquid spurts out when the can is opened ▪ Can is corroded, rusty, or leaky, swollen on top or bottom, dented on side seams ▪ Contents have unusual odors, a foamy or mushy texture ▪ Stored at temperatures above 100ºF or allowed to freeze and thaw
Fresh fruits and vegetables	▪ Unwashed, moldy
Cereal products, flour	▪ Moldy, infested with insects

*Home-canned foods should never be used in family day care, congregate meal sites, or other food service operations.

Possible Pathways of Pesticide Residues to a Fast-Food Meal

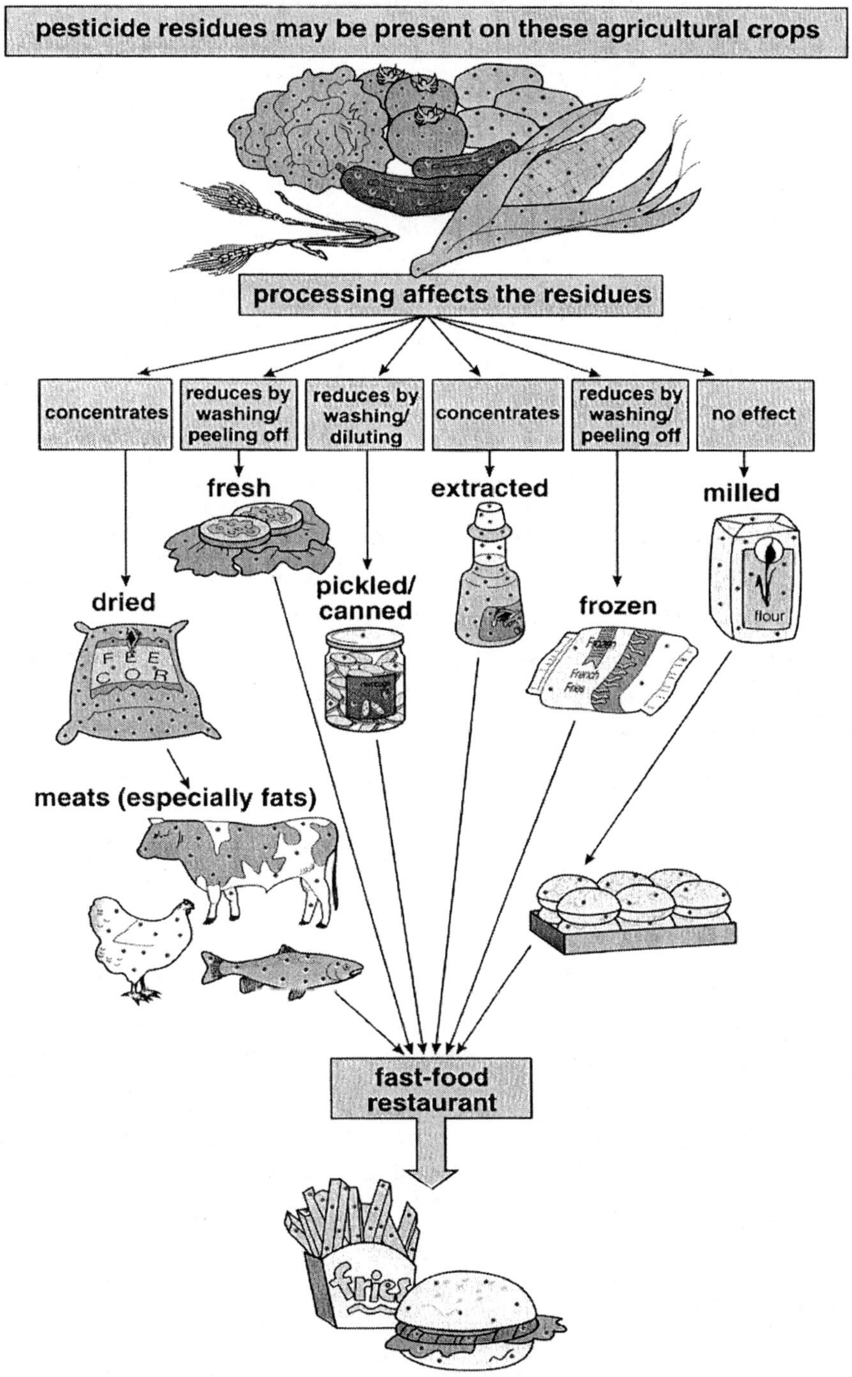

© 2000 Wadsworth/Thomson Learning

PRACTICE TEST

The following items will help you evaluate your understanding of this lesson. Use the answer key at the end of the lesson to check your answers or to locate material related in each question.

Multiple Choice

Select the one choice that best answers the question.

1. Which of the following is the major food source for transmission of *Campylobacter jejuni*?
   A. Raw poultry
   B. Uncooked seafood
   C. Contaminated water
   D. Imported soft cheeses

2. Which of the following hazards in our food supply is of most concern according to the Food and Drug Administration?
   A. Natural toxins
   B. Food additives
   C. Microbial food poisoning
   D. Environmental contaminants

3. What organism is the most *deadly* cause of food-borne illness in the United States?
   A. *Salmonella*
   B. *Clostridium botulinum*
   C. *Staphylococcus aureus*
   D. *Listeria monocytogenes*

4. Which of the following would most likely result from placing cooked hamburger patties on the same plate that held the uncooked patties?
   A. Flavor declination
   B. Meat juice retention
   C. Fat drippings exudation
   D. Microbial cross-contamination

5. Which of the following is the largest single group of food additives?
   A. Antioxidants
   B. Coloring agents
   C. Antimicrobial agents
   D. Flavoring agents

6. All of the following practices are known to minimize exposure to food pesticide residues EXCEPT __________.
   A. throwing away the outer leaves of leafy vegetables
   B. using a knife to peel citrus fruits rather than biting into the peel
   C. throwing away the fats and oils in broths and pan drippings from cooked meats
   D. washing waxed fruits and vegetables in water to remove the wax-impregnated pesticides

7. Which of the following is (are) the best known and most widely used antimicrobial agents?
   A. Salt
   B. Nitrites
   C. Sugar
   D. A and C

8. Monosodium glutamate __________.
   A. is used widely in Asian restaurants
   B. is an example of a flavor enhancer
   C. can be added to foods for infants
   D. A and B

9. What substance is commonly added to public water supplies to disinfect the water?
   A. Ozone
   B. Flouride
   C. Chlorine
   D. Penicillin

10. Why have bottled water sales increased dramatically over the past several years?
    A. People believe that tap water from the public water supply is unhealthy or unclean.
    B. Manufacturers of bottled water have made it inexpensive to buy when compared to tap water.
    C. Grocery stores have increased the shelf space, making bottled water more visible to consumers.
    D. All of the above.

11. An example of an insect that can be used to repel or eat other insects that damage or destroy crops is the __________.
    A. cockroach
    B. June bug
    C. ladybug
    D. earthworm

12. In order to avoid the side effects of MSG or other food additives, people should __________.
    A. learn to read labels to find the offensive additive
    B. eat more fresh fruit, vegetables, and whole grains
    C. ask that the offensive food additive be left out of foods when eating out
    D. all of the above

13. Time and temperature are two components of safe food handling. The third basic component is __________.
    A. speed
    B. cleanliness
    C. frequency
    D. intensity

14. When handling fruits, vegetables, and meats in the kitchen, the best advice is to __________.
    A. avoid cross-contamination by sanitizing utensils
    B. keep foods refrigerated until ready to eat, then serve immediately
    C. avoid taking a bite of raw food unless it is stamped “Safe For Consumption”
    D. wash hands and face immediately after handling meats

Fill in the Blank

Insert the correct word or words in the blank for each item.

15. Sales of soft drinks, beer, and wine have taken a backseat compared to sales of ________________________.

16. One alternative to pesticide use in organic farming is ____________________.

Essay Questions

Reflect on the items listed below; one or more may be included on the exam.

17. Describe three methods of eliminating microbes on kitchen cutting boards and utensils, and identify the advantages and disadvantages of each.

18. Can consumers concerned about food contamination eliminate all poisons from their diet by eating only "natural" foods? Why or why not? What would you recommend to avoid toxicity from natural food constituents?

ANSWER KEY

The following provides the answers and references for the practice test questions.

	Answer	Learning Objectives	References
1.	A	LO 2	Sizer/Whitney, p. 436
2.	C	LO 1	Sizer/Whitney, p. 435
3.	B	LO 1	Sizer/Whitney, p. 438
4.	D	LO 2	Sizer/Whitney, p. 441
5.	D	LO 5	Sizer/Whitney, p. 463
6.	D	LO 3	Sizer/Whitney, p. 452
7.	D	LO 2	Sizer/Whitney, p. 464
8.	D	LO 5	Sizer/Whitney, p. 466
9.	C	LO 6	Video
10.	A	LO 6	Video
11.	C	LO 4	Video
12.	D	LO 5	Video
13.	B	LO 2	Video
14.	A	LO 2	Video
15.	bottled water	LO 6	Video
16.	Answer may include any of the following: Earthworm castings Bat guano Seedless cow manure Ladybugs Compatible plants	LO 4	Video
17.		LO 2	Sizer/Whitney, pp. 440–443
18.		LO 4	Sizer/Whitney, pp. 472–480

Lesson 26

Applied Nutrition

OVERVIEW

This lesson concludes the twelve-month travels of the three *Nutrition Pathways* subjects profiled throughout this telecourse. You observed them as they were evaluated by nutrition or other allied health professionals; you watched them as they traveled down their respective pathways to healthier lives; and you examined the choices they made and the challenges they faced along the way. In this lesson, you will see them interact with each other and with the professionals that assisted them in their journey in a talk show format hosted by Debra Duncan of the ABC Network affiliate in Dallas, Texas. The host will ask the subjects to recap their experiences and to explain the difficulties they faced and the successes they had as they attempted to change their lives for the better. To conclude, each subject will offer advice to help others achieve success in implementing personal nutrition and lifestyle changes to improve health and well-being.

LEARNING OBJECTIVES

Upon completing this lesson, you should be able to:

1. Explain why the three *Nutrition Pathways* subjects chose to participate in the program.
2. Explain the reasons why some people fail when attempting to change nutrition and lifestyle habits.
3. Describe the most significant outcomes of the program for each subject.
4. Summarize the nutrition philosophy associated with this program.
5. State the importance of exercise in achieving success with regard to improvements in health and well-being.
6. State the advice you would give to someone who wants or needs to change nutrition or lifestyle habits.

LESSON ASSIGNMENT

Review the following assignment in order to schedule your time appropriately. Pay careful attention; the titles and numbers of the textbook chapter, the telecourse guide lesson, and the video program may be different from one another.

Text: There is no textbook reading assignment for this lesson.

Video: "Applied Nutrition" from the series *Nutrition Pathways*

LESSON FOCUS POINTS

The Focus Points for this lesson are identical to the Learning Objectives.

PRACTICE TEST

The following items will help you evaluate your understanding of this lesson. Use the answer key at the end of the lesson to check your answers or to locate material related in each question.

Multiple Choice

Select the one choice that best answers the question.

1. The reasons the Pathways subjects chose to participate in the *Nutrition Pathways* program included all of the following EXCEPT __________.
   A. a family history of heart disease
   B. a family history of diabetes
   C. having colon cancer
   D. being overweight

2. The goal of the weight-loss subject to lose 40 pounds in one year was for her __________.
   A. realistic
   B. unrealistic
   C. achieved
   D. B and C

3. Among the reasons why people fail in changing nutrition and lifestyle habits include __________.
   A. setting goals too high
   B. changing old habits too slowly
   C. expecting too little from the efforts that go into changing old habits
   D. all of the above

4. The greatest success for the subject who has type 2 diabetes involved _______.
   A. losing 50 pounds by the end of the year
   B. jogging 2 miles in twenty minutes
   C. going from two insulin injections per day to none
   D. all of the above

5. The philosophy used in *Nutrition Pathways* can be best summed up in which one of the following statements?
   A. "Change your life."
   B. "Balance your choices."
   C. "Deprivation breeds success."
   D. "Variety equals challenge."

Essay Questions

Reflect on the items listed below; one or more may be included on the exam.

6. Describe the importance of exercise with respect to stress reduction and balance in people's lives and its impact on type 2 diabetes, high blood cholesterol, and overweight.

7. What advice would you give to someone who wanted or needed to change nutrition or lifestyle habits? Include comments on the following: the need to evaluate current physical and/or nutrition status; steps to improve nutrient status or physical status; ways to handle challenges that might interfere with healthful choices; and the importance of balance and moderation in choices.

ANSWER KEY

The following provides the answers and references for the practice test questions.

	Answer	Learning Objectives	References
1.	C	LO 1	Video
2.	B	LO 2	Video
3.	A	LO 2	Video
4.	C	LO 3	Video
5.	B	LO 4	Video
6.		LO 5	Video
7.		LO 6	Video

Contributors

We gratefully acknowledge the valuable contributions to this course from the following individuals. The titles listed were accurate when the video programs were recorded, but may have changed since the original taping.

LESSON 1—"NUTRITION BASICS"

Jeanne P. Goldberg, Ph.D., R.D., Professor, Tufts University, Boston, MA

LESSON 2—"THE DIGESTIVE SYSTEM"

Judy Suneson, R.D., L.D., Clinical Dietitian, Baylor University Medical Center, Dallas, TX

LESSON 3—"CARBOHYDRATES: SIMPLE AND COMPLEX"

Judith Hallfrisch, Ph.D., Research Leader for the Metabolism and Nutrient Interactions Lab, Beltsville Human Nutrition Research Center, Beltsville, MD

LESSON 4—"CARBOHYDRATES: FIBER"

James Anderson, Ph.D., Professor of Medicine and Clinical Nutrition, University of Kentucky, Lexington, KY
Mary Sue Cole, R.D., L.D., County Extension Agent, Greenville, TX

LESSON 5—"FATS: THE LIPID FAMILY"

Gloria Vega, Ph.D., Professor of Clinical Nutrition, University of Texas Southwestern Medical Center, Dallas, TX
Clay King, Ph.D., Associate Professor of Nutrition and Food Sciences, Texas Woman's University, Denton, TX

LESSON 6—"FATS: HEALTH EFFECTS"

Jo Ann Carson, M.S., R.D., L.D., Director of Clinical Dietetics Program, University of Texas Southwestern Medical Center, Dallas, TX

LESSON 7—"PROTEIN: FORM AND FUNCTION"

Martha Groblewski, *Ph.D., R.D.*, College of Health Science, Old Dominion University, Norfolk, VA

LESSON 8—"THE PROTEIN CONTINUUM"

Cecile H. Edwards, Ph.D., L.N., C.N.S., Professor of Nutrition, Howard University, Washington, DC
John A. McDougall, M.D., Saint Helena Hospital, Santa Rosa, CA

LESSON 9—"METABOLISM"

Rene Frenkle, Ph.D., Biochemist at the University of Texas Southwestern Medical Center, Dallas, TX
Do-Hyun Choe, O.M.D., Doctor of Oriental Medicine, Dallas, TX
Dale Swanholm, M.D., F.A.A.F.P, Board Certified Physician in Family Practice and Geriatrics, Dallas, TX

LESSON 10—"WEIGHT CONTROL: ENERGY REGULATION"

Craig Johnson, Ph.D., Director of Eating Disorders, Laureate Psychiatric Hospital, Clinical Psychology Professor, University of Tulsa, Tulsa, OK
Judith S. Stern, Ph.D., Professor of Nutrition and Internal Medicine at the University of California at Davis, Davis, CA

LESSON 11—"WEIGHT CONTROL: HEALTH EFFECTS"

Kelly D. Brownell, Ph.D., Department of Psychology, Yale University, New Haven, CT

LESSON 12—"VITAMINS: WATER SOLUBLE"

Jethon' Sharrieff, Associate Store Director, Whole Foods Market, Chicago, IL
Lillie R. Williams, Ph.D., R.D., Chairperson of the Department of Nutritional Sciences, Howard University, Washington, DC

LESSON 13—"VITAMINS: FAT SOLUBLE"

Ishwarlal Jialal, Ph.D., Associate Professor of Medicine and Pathology, University of Texas Southwestern Medical Center, Dallas, TX

LESSON 14—"MAJOR MINERALS AND WATER"

Claudia A. Barner, Ph.D., R.D., L.D., University of Texas Southwestern Medical School, Dallas, TX
K. Sakhaee, M.D., Professor of Internal Medicine, University of Texas Southwestern Medical Center for Mineral Metabolism and Clinical Research, Dallas, TX

LESSON 15—"TRACE MINERALS"

Richard Anderson, M.D., Program Director, Human Nutrition Research Center/USDA, Beltsville, MD
Rebecca Costello, Ph.D., Deputy Director, Office of Dietary Supplements, National Institute of Health, Bethesda, MD
Melissa Feagley, Clinical Dietitian, Presbyterian Hospital of Plano, Plano, TX
Barbara Gollman, Ph.D., R.D., Registered Dietitian and Nutrition Consultant, Dallas, TX
Jeanette Hasse, Ph.D., R.D., Transplant Nutrition Specialist, Baylor University Medical Center, Dallas, TX
Jay Schulman, D.D.S., Dentist and Professor, Baylor College of Dentistry, Dallas, TX
Ann Zogbaum, M.S., R.D., Oncology Dietitian, Hartford Hospital, Hartford, CT

LESSON 16—"PHYSICAL ACTIVITY: FITNESS BASICS"

John Duncan, Ph.D., Chief of Clinical Application, Cooper Institute for Aerobics Research, Dallas, TX

LESSON 17—"PHYSICAL ACTIVITY: BEYOND FITNESS"

Nancy Clark, M.S., R.D., Sports Medicine Brookline, Boston, MA
Melvin H. Williams, Ph.D., Eminent Professor of Health, Physical Education, and Recreation, Old Dominion University, Norfolk, VA

LESSON 18—"LIFE CYCLE: PREGNANCY"

Erica Gunderson, M.S., M.Ed., R.D., Doctoral Program in Epidemiology, University of California at Berkeley, Berkeley, CA

LESSON 19—"LIFE CYCLE: LACTATION AND INFANCY"

Kathryn G. Dewey, Ph.D., Professor in the Department of Nutrition, University of California at Davis, Davis, CA

LESSON 20—"LIFE CYCLE: CHILDHOOD AND ADOLESCENCE"

Craig Johnson, Ph.D., Director of Eating Disorders, Laureate Psychiatric Hospital, Clinical Psychology Professor, University of Tulsa, Tulsa, OK
Laurell Mellin, M.A., R.D., Director of the Center for Child and Adolescent Obesity, Associate Clinical Professor of Family Medicine and Pediatrics, University of California at San Francisco, San Francisco, CA

LESSON 21—"LIFE CYCLE: ADULTHOOD AND AGING"

Floristene Johnson, R.D., Deputy Regional Administrator for the Administration on Aging, Dallas, TX
Robert M. Russell, Ph.D., Associate Director of the Nutrition Research Center on Aging, Professor of Medicine and Nutrition, Tufts University, Boston, MA

LESSON 22—"DIET AND HEALTH: CARDIOVASCULAR DISEASE"

Donna Israel, Ph.D., R.D., L.D., L.P.C., President of Preferred Nutrition Therapist Incorporated, Owner of the Fitness Formula, Dallas, TX
Emily Malorzo, Registered Dietitian, Baylor Heart and Vascular Hospital, Dallas, TX
Ernst J. Schaefer, M.D., Senior Scientist and Director, Lipid Metabolism laboratory, Tufts University, Boston, MA

LESSON 23—"DIET AND HEALTH: CANCER, IMMUNOLOGY, AND AIDS"

Stacey J. Bell, D.Sc., R.D., Research Dietitian, New England Deaconess Hospital, Boston, MA
Lauri Wright, R.D., James A. Haley Veterans Hospital, Tampa, FL
Faith D. Ottery, M.D., Ph.D., Surgical Oncologist, Fox Chase Cancer Center, Philadelphia, PA

LESSON 24—"DIET AND HEALTH: DIABETES"

Susan L. Thom, R.D., L.D., C.D.E., 1994–1995 President of American Association of Diabetes Educators, Private Practice, Diabetes Associates, Cleveland, OH

Zaven H. Chakmakjian, M.D., Diplomate of Internal Medicine and Endocrinology/Metabolism, Associate Clinical Professor, University of Texas Southwestern Medical School, Dallas, TX

LESSON 25—"CONSUMER CONCERNS AND FOOD SAFETY"

Michael Jacobson, Ph.D., Founder and Executive Director for the Center for Science in the Public Interest, Washington, DC

Jethon' Sharrieff, Associate Store Director, Whole Foods Market, Chicago, IL

LESSON 26—"APPLIED NUTRITION"

Debbie Childers, R.D., L.D., C.D.E., Diabetes Nutrition Educator, Baylor University Medical Center, Dallas, TX

Donna Israel, Ph.D., R.D., L.D., L.P.C., President of Preferred Nutrition Therapist Incorporated, Owner of the Fitness Formula, Dallas, TX

Personal Dietary Analysis Project

ATTENTION: Read these instructions carefully BEFORE you begin!

Do NOT change your eating habits for this assignment. Your grade will depend on how well you analyze your eating habits, not on the foods you eat! The more carefully you complete your food diary, the more accurate your information will be.

Diet Analysis Plus software is available from the textbook publisher. Input your food records using the software to generate the required printouts. The users' guide for the computerized dietary analysis program has samples of what you will need to print.

PART 1—Keep Daily Food Records and Generate Computer Printouts:

I. Write down all the foods and beverages you eat over a three-day period, including ONE weekend day. Record the amount and a description of the food/beverage on the attached Daily Food Record (DFR) form.
   A. Use "ounces, cups, each, teaspoon, tablespoon" etc.
   B. Estimate serving sizes, weigh the food, and/or use food labels.
   C. A three-ounce portion of meat is about the size of a woman's palm or a deck of cards; a restaurant serving of vegetables is 1/2 to 2/3 cup.
   D. If unsure of serving size, ask the server or estimate on the large side.

II. Print out the following items and staple them together in the order listed below.
   A. Your personal profile (Recommended Daily Nutrients) based on your gender, activity level, height, weight, and age
   B. Your three days of foods/beverages including:
      - the foodlist for each day
      - ratios and percents for each day
      - a bar graph for each day
      - the Food Pyramid for each day
   C. A three-day *average* of all nutrients, including:
      - foodlist (some software does not provide a foodlist for the average)
      - ratios and percents
      - bar graph
      - Food Pyramid
   D. Daily Food Record forms (optional)

—Continued on next page

III. Submit the above printouts to your professor by the established deadline.

PART 2—Answer Questions to Analyze Your Eating Habits:

I. Interpret your results. Refer to the printouts of your three-day average to obtain data regarding your personal dietary intake. Use the textbook, *Nutrition: Concepts and Controversies*, 10th ed., to answer other nutrition-related questions. (Questions begin at the back of the Telecourse Guide, pp. 277–289.)

II. Type or print (neatly) all responses in the spaces provided on the question forms.

III. Submit the following to your professor by the established deadline. Staple pages together or put in a three-ring folder (no hard covers!).

- three-day *average* with foodlist (if available), ratios and percents, bar graph, and Food Pyramid from PART 1 above
- questions with answers

Daily Food Record

NAME: ______________________________ SS #: ______________________________

Write down ALL foods and beverages consumed throughout the day. Be as specific as possible about the *amount* as well as the *description* of the food item itself. For example, specify the type of bread, such as whole wheat, white, or rye; type of cheese, such as cheddar, Swiss, or American; whether fruits or vegetables are fresh, frozen, or canned; and if meats are lean only, or lean with some fat. The most accurate entries for *amounts* are for weighed foods in ounces (oz), grams (g), or pounds (lb). Other measures are cups (c), teaspoons (tsp), tablespoons (tbl), "each" or "piece." Weights of foods should be for edible portion only.

Try to maintain your normal eating patterns. Do not exclude foods or beverages that you perceive to be "bad" such as alcohol, candy, etc. You will gain more valuable information about your diet if you honestly analyze your eating habits.

Daily Food Record

Day #1 (Date: ______________)

AMOUNT:	FOOD AND BEVERAGE DESCRIPTION:

Daily Food Record

Day #2 (Date: ______________)

AMOUNT:	FOOD AND BEVERAGE DESCRIPTION:

Daily Food Record

Day #3 (Date: ______________)

AMOUNT:	FOOD AND BEVERAGE DESCRIPTION:

Carbohydrate and Fiber Intakes

DIRECTIONS: Refer to your three-day average printouts to answer the following questions. Type or print answers in the space provided.

1. ____________ What is your average percentage of total calories from carbohydrates? (See Source of Calories.)

2. ____________ Dietary guidelines state that 50 percent or more of your total calories should come from carbohydrates. How did your intake compare to the recommendation?

3. ____________ It is estimated that you should have 125 grams per day or more of carbohydrate. How many grams per day did you consume on the average? (See Value on Bar Graph.)

4. ____________ On the average, what percentage of your personal Recommended Daily Nutrients (RDN) did you consume in carbohydrates? (See Goal % on Bar Graph.)

5. ____________ The recommended intake of fiber per day is 20 to 35 grams. How many grams per day did you consume on the average? (See Value on Bar Graph.)

6. ____________ On the average, what percentage of your RDN did you consume in fiber? (See Goal % on Bar Graph.)

7. List all of your refined sugar sources; that is, colas, candy, pastries, and so forth for each day of your three-day intake.

   DAY 1: ____

   DAY 2: ____

   DAY 3: ____

—Continued on next page

8. It is recommended that less than 10 percent of total calories come from sweets. Estimate how your intake compared to the recommendation.

____

____

9. List three of your personal best sources of complex carbohydrates or fiber in your diet.

____

____

____

10. List three additional foods you could include in your diet to increase your intake of complex carbohydrates or fiber.

____

____

____

11. How can increasing your intake of complex carbohydrates/fiber and reducing your intake of concentrated sweets improve your health status? Cite page numbers in your text to support your answer.

Fat and Dietary Cholesterol Intakes

DIRECTIONS: Refer to your three-day average printouts to answer the following questions. Type or print answers in the space provided.

1. ______________ What is your average percentage of total calories from fat? (Source of Calories)

2. ______________ The dietary goal for total fat intake is less than 30 percent of total calories. How did your fat intake compare to the stated goal?

3. ______________ How many total grams of fat did you consume on an average day? (See Value on Bar Graph.)

4. ______________ On the average, what percentage of your personal RDN did you consume in fat? (See Goal % on Bar Graph.)

5. ______________ What percent of your calories came from saturated fat? (See Source of Fat.)

6. ______________ What percent of your calories came from monounsaturated fat?

7. ______________ What percent of your calories came from polyunsaturated fat?

8. ______________ How many milligrams of cholesterol did you consume on an average day? (See Value on Bar Graph.)

9. ______________ The recommended intake of dietary cholesterol is less than 300 mg per day. How did your intake compare to the recommendation?

—Continued on next page

10. Health recommendations state that less than 30 percent of total calories should come from fat in the following categories: less than 10 percent of fat calories from saturated fat; more than 10 percent of fat calories from monounsaturated fat; and about 10 percent of fat calories from polyunsaturated fat. How did your fat intake compare to the recommended categories?

____

____

11. List one food containing cholesterol from each day of your three days that you could limit to reduce total cholesterol intake.

____

____

____

12. List three foods in your diet you could limit to reduce total fat intake.

____

____

____

13. How can health be improved by decreasing total fat, saturated fat, and cholesterol, and increasing monounsaturated fat? Cite page numbers in your text to support answers.

____

____

____

____

____

____

____

____

____

Protein Intakes

DIRECTIONS: Refer to your three-day average printouts to answer the following questions. Type or print answers in the space provided.

1. ______________ What is the average percentage of your total calories that comes from protein? (Source of Calories)

2. ______________ Dietary guidelines recommend that protein should provide about 10 to 20 percent of the total intake of calories. How did your intake compare to the recommendation?

3. ______________ How many grams of protein did you consume on the average? (Bar Graph)

4. ______________ On the average, what percentage of your personal RDN did you consume in protein? (Goal %)

5. List one meat source of protein from each day of your three-day intake. For each meat source of protein, list combined plant sources of protein that you could eat in place of it.

Meat Source:	Plant Combination:
Day 1: ____________________	Day 1: ____________________
Day 2: ____________________	Day 2: ____________________
Day 3: ____________________	Day 3: ____________________

—Continued on next page

For each of the following questions, cite page numbers in your text to support your answer.

6. How would eating plant protein versus meat protein impact saturated fat, total fat, and cholesterol intakes?

7. How would the use of plant protein in place of meat protein affect the protein quality of the diet?

8. State the health concerns surrounding high-protein diets from animal sources.

Weight Control

DIRECTIONS: Refer to your three-day average printouts to answer the following questions. Type or print answers in the space provided.

1. ______________ What was your average calorie intake? (Bar Graph)

2. ______________ What percentage of your personal RDN did you consume? (Goal %)

3. ______________ If you consumed the same amount of calories per day over a period of time (see #2 above), would you lose, gain, or maintain your weight?

4. ______________ Is your personal goal to lose, gain, or maintain your weight?

For questions 5 through 9 cite page numbers in your text to support your answers.

5. What should be the primary consideration when setting a weight goal?
6. How would consuming fewer than 1,200 calories per day impact metabolism?
7. How would consuming fewer than 1,200 calories per day impact health and nutrition status?
8. In addition to calorie control, list three other factors that should be included in a successful weight-control program.
   A.
   B.
   C.
9. What two physical activities could you perform to help you reach your goal weight?
   A.
   B.
   C.

Vitamin Intakes

DIRECTIONS: Refer to your three-day average printouts to answer the following questions. Type or print answers in the space provided.

1. Look at your average intake of vitamins. List the water-soluble vitamins that were *less* than 75 percent of your personal RDN. (See Goal %)

2. List the water-soluble vitamins that were *greater* than 75 percent of your personal RDN.

3. List the fat-soluble vitamins that were *less* than 75 percent of your personal RDN.

4. List the fat-soluble vitamins that were *greater* than 75 percent of your personal RDN.

5. List one good food source for each of the water-soluble vitamins.

6. List one good food source for each of the fat-soluble vitamins.

—Continued on next page

For questions 7 through 9, cite page numbers in your text to support your answers.

7. Explain why excess water-soluble vitamin intakes are *less* of a health concern than excess fat-soluble vitamin intakes.

8. State the health concerns surrounding vitamin supplements.

9. What are the health concerns surrounding antioxidant vitamins?

Mineral Intakes

DIRECTIONS: Refer to your three-day average printouts to answer the following questions. Type or print answers in the space provided.

1. Look at your average intake of minerals. List the *major* minerals that were *less* than 75 percent of your personal RDN. (Goal %)

2. List the *major* minerals that were *greater* than 75 percent of your personal RDN.

3. List the *trace* minerals that were *less* than 75 percent of your personal RDN.

4. List the *trace* minerals that were *greater* than 75 percent of your personal RDN?

5. List one good food source for each of the *major* minerals.

6. List one good food source for each of the *trace* minerals.

—*Continued on next page*

MINERAL INTAKES—*Continued*

For questions 7 and 8, cite page numbers in your text to support your answers.

7. Health recommendations state that sodium intake should be limited to 2,400 milligrams per day.

   A. How did your intake of sodium compare to the recommendation?

   B. How could more than 2,400 milligrams per day of sodium adversely affect health?

8. How could excessive intakes of iron adversely affect health?

The Food Pyramid and Other Information

DIRECTIONS: Refer to your three-day average printouts to answer the following questions. Type or print answers in the space provided.

The Food Pyramid:

1. What are the recommended servings for each group on the Food Pyramid?

   ____

   ____

   ____

   ____

2. Examine your *average* Food Pyramid servings for each food group, then respond to the following:

   A. What food groups were in the preferred range of servings?

   ____

   ____

   ____

   ____

   B. What food groups were in the less-than-preferred range of servings?

   ____

   ____

   ____

   ____

—Continued on next page

3. Why is it important to eat the recommended number of servings per day for each level of the Food Pyramid? Cite page numbers in your text to support your answer.

____

____

____

____

Other Information:

4. Consider all of the information you obtained from analyzing your food intake and from reading the text. What general modifications or changes could you make to improve your current eating habits and current lifestyle choices?

____

____

____

____

5. What difficulties or challenges must you face before you can implement the above changes?

____

____

____

____